Acoustic Emission - New Perspectives and Applications

Edited by Mahmut Reyhanoglu

Published in London, United Kingdom

IntechOpen

Supporting open minds since 2005

Acoustic Emission - New Perspectives and Applications
http://dx.doi.org/10.5772/intechopen.95657
Edited by Mahmut Reyhanoglu

Contributors
Behzad Behnia, Vasishta Bhargava Nukala, Satya Prasad Maddula, Swamy Naidu Venkata Neigapula, Rahul Samala, Chinmaya Prasad Padhy, Gaurav Sharma, Shruti Sharma, Sandeep Kumar Sharma, Tanmoy Bose, Subhankar Roy, Jie Deng, Nansha Gao, Mahmut Reyhanoglu

First published in London, United Kingdom, 2022 by IntechOpen
IntechOpen is the global imprint of INTECHOPEN LIMITED, registered in England and Wales, registration number: 11086078, 5 Princes Gate Court, London, SW7 2QJ, United Kingdom
Printed in Croatia

British Library Cataloguing-in-Publication Data
A catalogue record for this book is available from the British Library

Additional hard and PDF copies can be obtained from orders@intechopen.com

Acoustic Emission - New Perspectives and Applications
Edited by Mahmut Reyhanoglu
p. cm.
Print ISBN 978-1-80355-105-0
Online ISBN 978-1-80355-106-7
eBook (PDF) ISBN 978-1-80355-107-4

For EU product safety concerns:
IN TECH d.o.o., Prolaz Marije Krucifikse Kozulić 3, 51000 Rijeka, Croatia, info@intechopen.com or visit our website at intechopen.com.

Meet the editor

Mahmut Reyhanoglu is currently Director of Robotics Engineering, Columbus State University (CSU), Georgia, USA. Prior to joining CSU, he was the Glaxo Wellcome Distinguished Professor of Engineering, University of North Carolina at Asheville, USA. His extensive research makes use of advanced mathematical techniques and models that arise from fundamental physical principles. His major research interests are in the areas of nonlinear dynamical systems and control theory, with particular emphasis on applications to mechanical and aerospace systems, robotics, and mechatronics. He has edited five books and authored/co-authored five book chapters and more than 140 peer-reviewed journal/proceedings papers. He served on the Transactions on Automatic Control Editorial Board and the Control Systems Society Conference Editorial Board of the Institute of Electrical and Electronics Engineers (IEEE). He also served as a member of the International Program Committee for several conferences and as a member of the Guidance, Navigation, and Control Technical Committee of the American Institute of Aeronautics and Astronautics (AIAA). He is currently an editor of the *International Journal of Aerospace Engineering* and *Electronics*.

Contents

Preface

Acoustic emission (AE) is a phenomenon involving sound wave generation by materials subject to deformation, fracture, and chemical reactions. The use of AE has been growing tremendously in process monitoring and quality control in manufacturing processes since its discovery in the early 1950s in Germany. Structural testing and assessment, process monitoring, and material characterization constitute three broad application areas of AE techniques. Quantitative and qualitative characteristics of AE waves have been studied widely in the literature. This book reviews major research developments over the past few years in application of AE in numerous fields, including aerospace, automotive, biomedical, manufacturing, and civil and materials engineering fields. It brings together important contributions from renowned international researchers to provide an excellent survey of new perspectives and paradigms of acoustic emissions. In particular, this book presents applications of AE in cracking and damage assessment in metal beams, asphalt pavements, and composite materials as well as studying noise mitigation in wind turbines and cylindrical shells.

Chapter 1 introduces the AE technique by providing a brief background and summarizing the application areas of AE including aerospace, automotive, biomedical, manufacturing, and civil and materials engineering fields. Chapter 2 is devoted to cracking and damage assessment in Steel-Reinforced Concrete (Steel-RC) beams and Glass Fiber-Reinforced Polymer–Reinforced Concrete (GFRP-RC) beams. The AE technique is shown to be an effective Non-Destructive Testing (NDT) tool for concrete structures. The behavior of cracks of the RC beams from loading to failure using an AE parameter analysis-based method is analyzed. The chapter provides a classification of the crack types and damage levels that occur in these RC beams with varying percentage tension reinforcement ratios. Chapter 3 focuses on various applications of the AE technique in evaluating low-temperature cracking in asphalt pavement materials including: (1) assessment of low-temperature cracking performance of asphalt binders and asphalt mixtures and (2) quantitative characterization of rejuvenators' efficiency in restoring aged asphalt pavements to their crack-resistant state. The AE-based embrittlement temperature results of twenty-four different asphalt materials consisting of eight different binders, each at three oxidative aging levels, are presented. Results show that the AE-based embrittlement temperatures were consistently lower than the Bending Beam Rheometer (BBR-based) critical cracking temperatures. Chapter 4 deals with polymer matrix-based composites that are widely used for various applications in aerospace, automobile, marine, sports, construction, and electrical industries. Complicated defects like delamination present in the composite laminates can be detected effectively using nonlinear acoustic wave spectroscopy (NAWS). One NAWS technique for detecting delamination is based on intensification of vibration amplitudes at the delamination location, known as the local defect resonance (LDR) technique. This technique is quite established in early detection of damages in composite structures. In this chapter, a numerical investigation for detecting delamination in GFRP composite based on the vibro- thermography technique is discussed. A single periodic LDR frequency excitation is used to excite the GFRP plate, resulting in a local temperature rise at delamination region due to frictional heating at the damage interface. An explicit dynamic temperature displacement analysis is carried out for a specific period of LDR excitation. Subsequently, a heat transfer analysis is performed to observe the temperature difference at the top surface of the delaminated GFRP plate. A numerical investigation is carried out based on LDR excitation for high-contrast imaging of delamination in composite materials using vibro-thermography. Chapter 5 presents an analysis of sound radiation from periodic Acoustic Black Hole (ABH) shells. It is shown that by reducing thickness according to the power law, the wave velocity is substantially lowered, while the wave number is increased when it propagates to the ABH center, where the damping dissipates the vibrational energy very efficiently. The focus is placed on reducing the sound emission from cylindrical shells via embedding periodic ABHs. Finally, Chapter 6 presents a computational analysis of trailing edge bluntness vortex shedding noise for a 2-MW horizontal axis wind turbine (HAWT) for trailing edge thicknesses of 0.1 % and 0.5 % local chord using the original Brooks–Pope–Marcolini (BPM) model. The original BPM and the modified BPM results for trailing edge bluntness noise are clearly presented.

Mahmut Reyhanoglu
Columbus State University,
United States of America

Chapter 1

Introductory Chapter: Acoustic Emission

Mahmut Reyhanoglu

1. Introduction

Acoustic emission (AE) is a relatively new non-destructive Testing (NDT) technique. Structural testing and assessment, material characterization, and process monitoring are three important application areas of AE. A comprehensive introduction to the AE technique can be found in [1].

The AE technique is one of the most reliable NDT techniques for detecting and monitoring damages and defects in different structures. AE has been effectively employed for fracture behavior monitoring and fatigue detection in various materials including composites, metals, concrete, fiberglass, ceramics, plastics, and wood. It has also been used for fault and pressure leak detection in pipes, tanks, and vessels.

There are several electronic instruments that can be used to digitize and store large numbers of high-speed digital waveform signals of AE. Common instruments used in AE include preamplifiers, amplifiers, filters, sensors and other data collection, analysis, and storage equipment such as computers, oscilloscopes, and voltmeters. Preamplifiers are used to amplify a weak signal and to reduce the interference from noise, while the piezoelectric sensors are used for the conversion of mechanical AE waves into electrical voltages. The overall objective of the measurement is to determine the various AE parameters such as the frequency range (controlled by filters) that exist in the system by observing and measuring the performance of AE amplifiers and sensors. These are very useful mechanisms for measuring the essential AE parameters such as event, count, energy moment, maximum amplitude, hit, energy, arrival-time difference, RMS (root mean square) voltage, rise time, spectrum, frequency, and duration [2].

2. Applications of AE techniques

Application areas of AE span numerous fields, including aerospace, automotive, biomedical, manufacturing, civil, and materials engineering fields.

Monitoring the condition and predicting the life of the main structures of an aircraft play significant role in guaranteeing the flight safety. AE techniques are successfully employed in damage and crack identification and monitoring in aircraft composite and steel structures [3, 4].

AE is extensively used in the automotive industry in fault diagnosis of internal combustion engines (ICEs). Reference [5] deals with advanced techniques based on vibro-acoustic signals that can diagnose and monitor ICE malfunctions under vehicle operating conditions. Reference [6] develops new AE models and effective wavelet-based AE signal processing techniques for monitoring lubrication conditions.

The primary use of AE technique in biomedical field is bone condition assessment under different loading conditions, in osteoporosis and in fracture healing process monitoring [7]. AE can be used for detecting defects in tissues and materials, predicting failure, and monitoring damage progression in real time [8].

AE has been used as a widely applied technique in manufacturing process monitoring due to its sensitivity to process parameters. The use of AE as a monitoring technique for machining operations comes with more advantages, one of which is its ability to detect machine vibrations from those of AE signals due to high-frequency range and sensitivity of AE signals, thus preventing it from interfering with the cutting operation [9, 10].

Structural health monitoring (SHM) in civil engineering involves AE technique for detecting cracks in structures. This technique relies on the high-frequency ultrasonic waves generating energy that is rapidly emitted from a material throughout from the initiation to growth progression of cracks. The wide applicability of AE technique is evident in several metal piping system evaluations and fiberglass-reinforced plastics (FRP) and concrete bridges [11, 12].

AE originates from stress waves generated as a result of the growth or movement that takes place in solid defects. When a composite material is subject to a mechanical load, it can experience matrix cracking, debonding, and delamination. AE is a powerful technique capable of detecting these damage types in composites [13, 14].

Author details

Mahmut Reyhanoglu
Columbus State University, Georgia, USA

*Address all correspondence to: mreyhanoglu@icloud.com

References

[1] Scruby CB. An introduction to acoustic emission. Journal of Physics E: Scientific Instruments. 1987;**20**(8):946

[2] Grosse CU, Ohtsu ME. Acoustic Emission Testing. Berlin, Heidelberg: Springer-Verlag; 2008

[3] Diamanti K, Soutis C. Structural health monitoring techniques for aircraft composite structures. Progress in Aerospace Sciences. 2010;**46**(8): 342-352

[4] Pullin R, Eaton MJ, Hensman JJ, Holford KM, Worden K, Evans S. Validation of acoustic emission (AE) crack detection in aerospace grade steel using digital image correlation. Applied Mechanics and Materials. 2010; **24**:221-226

[5] Delvecchio S, Bonfiglio P, Pompoli F. Vibro-acoustic condition monitoring of internal combustion engines: A critical review of existing techniques. Mechanical Systems and Signal Processing. 2018;**99**:661-683

[6] Wei N, Gu JX, Gu F, Chen Z, Li G, Wang T, et al. An investigation into the acoustic emissions of internal combustion engines with modelling and wavelet package analysis for monitoring lubrication conditions. Energies. 2019;**12**:640-659

[7] Shrivastava S, Prakash R. Assessment of bone condition by acoustic emission technique: A review. Journal of Biomedical Science and Engineering. 2009;**2**(03):144

[8] Kohn DH. Acoustic emission and nondestructive evaluation of biomaterials and tissues. Critical Reviews in Biomedical Engineering. 1995;**23**(3-4):221-306

[9] Govekar E, Gradisek J, Grabec I. Analysis of acoustic emission signals and monitoring of machining processes. Ultrasonics. 2000;**38**(1-8):598-603

[10] Jemielniak K, Arrazola PJ. Application of AE and cutting force signals in tool condition monitoring in micro-milling. CIRP Journal of Manufacturing Science and Technology. 2008;**1**(2):97-102

[11] Grosse CU, Reinhardt HW, Finck F. Signal-based acoustic emission techniques in civil engineering. Journal of Materials in Civil Engineering. 2003;**15**(3):274-279

[12] Ai Q, Liu CX, Chen XR, He P, Wang Y. Acoustic emission of fatigue crack in pressure pipe under cyclic pressure. Nuclear Engineering and Design. 2010;**240**(10):3616-3620

[13] Ohtsu M. The history and development of acoustic emission in concrete engineering. Magazine of Concrete Research. 1996;**48**(177): 321-330

[14] Wevers M. Listening to the sound of materials: Acoustic emission for the analysis of material behavior. NDT and E International. 1997;**30**(2):99-106

Chapter 2

Crack Classification in Steel-RC and GFRP-RC Beams with Varying Reinforcement Ratio Using AE Parameters

Gaurav Sharma, Shruti Sharma and Sandeep Kumar Sharma

Abstract

The main aim of this chapter is to monitor the cracking and damage assessment in steel-reinforced concrete (steel-RC) and glass fibre polymer-reinforced concrete (GFRP-RC) beams along with varying percentages of tension reinforcement ratio. Beam specimens measuring (150 × 230 × 2100) mm were tested using a four-point bending flexural test using a universal testing machine together with an AE monitoring system. Acoustic emission (AE) has been applied for the early monitoring of steel-RC and GFRP-RC structures using AE parameters such as cumulative AE hits, average frequency, rise angle, amplitude, duration and AE XY plots to evaluate the micro and macro cracking in the steel-RC and GFRP-RC beams specimens. The most popular applications of AE signal in structural health monitoring are specified on crack monitoring, quantifying the degree of damage, and crack classification. In this research, the results indicated that the average frequency and rise angle parameter of AE signal are applied to classify the types of cracks (flexural or shear cracks) that occur in steel-RC and GFRP-RC beams along with varying percentages of tension reinforcement ratio subjected to flexural loading. As a result of these findings, the AE approach may be used to examine crack monitoring and crack classification in steel and GFRP-RC beams with different percentages of tension reinforcement ratios.

Keywords: load-deflection, AE, crack classification, hits, GFRP bars, steel bars

1. Introduction

The most common problem associated with coastal infrastructure in metro cities like Mumbai, Bangalore, and Chennai is corrosion, which leads to cracking (micro-macro) and resulting in gradual ageing of the structure and its components, as a result of climate change and sea-level rises [1]. Moreover, the National Association of Corrosion Engineers (NACE) has detailed the adverse impact on the Indian economy for the corrosion of reinforcement in structural components about U.S. $26.1 billion (2.4% of the nations).

Gross domestic products (GDP) is spent annually for the corrosion of infrastructure by the government of India [2]. Numerous strategies have been proposed to postpone corrosion in reinforced concrete structures to minimise these huge expenditures. Commonly suggested methods include the use of stainless steel [3] and epoxy-coated rebars [4] in place of conventional steel bars, admixing corrosion inhibitors [5], self-healing compounds in concrete [6], and use of polymer concrete. It is important to note that these methods only delay the onset of corrosion and do not prevent it entirely. Recently some researchers have also suggested the use of self-healing micro-capsules for corrosion protection of metal [6, 7]. Apart from that, these specific methods have scalability issues in structural applications and are not cost-effective. Steel corrosion is a major problem in the construction industry, and several approaches have been tried to combat it, but they have proven to be either expensive or ineffective. Civil engineers from all over the world are challenged and in search of new and non-corroded affordable construction materials as well as innovative approaches and systems to problem-solving [8].

In many civil engineering applications, fibre-reinforced polymer (FRP) composites reinforcement has been introduced as an alternative or substitute material that can replace traditional reinforcing steel [9–11]. Apart from all this, FRP is a non-corrosive material consisting of a polymer matrix reinforced with fibres [12]. The fibres are usually aramid, basalt, carbon, and glass, although other fibres such as asbestos or paper, or wood have been sometimes used. On the other hand, the polymer is usually an epoxy, vinyl ester, or polyester thermosetting plastic and phenol-formaldehyde resins are still in use. FRPs are commonly used in the aerospace, automotive, marine, and construction industries. Glass fibre polymer (GFRP) bars have been predominantly suggested for engineering applications vis-à-vis economy and specific strength properties among other fibres [13].

Being non-corrosive, GFRP composite bars has many advantages such as high strength-to-weight ratio, electromagnetic-neutrality, light-weight, ease of handling, high longitudinal tensile strength, and non-magnetic characteristics, are easily constructed, and can be tailored to satisfy performance requirements [14, 15]. FRP composites have been used as internal reinforcement in concrete, bridge decks, modular structures, formwork, and external reinforcement for strengthening and seismic upgrading in modern construction and restoration of structures due to their favourable qualities. GFRP bars can be utilised in place of steel rebars in harsh exposure situations such as coastal settings, as well as in a variety of other structural applications such as wharves, box culverts, dry docks, and retaining walls [16]. Furthermore, due to their linearly elastic stress-strain relationship up to failure, GFRP reinforcing bars react differently from typical steel reinforcing bars. Furthermore, as compared to steel-reinforced concrete members, the lower modulus of elasticity of GFRP reinforcing bars produces a significant drop in flexural stiffness of GFRP-RC members after cracking and, as a result, greater deflection/deformations under service or loading circumstances [17, 18].

As a result, the serviceability limit state is frequently used to guide the design of GFRP-RC flexure members. In addition, relevant design codes and guidelines for the use of GFRP bars in RC structures have been developed [8]. More research is still needed to provide the required confidence through a better understanding of the flexural behaviour of GFRP-RC [8] using the non-destructive acoustic emission (AE) technique. The AE technique (AET) is considered as one of the most promising techniques from various types of Non Destructive Testing (NDT) methods [19]. The AE method and other NDT methods differ in two main features. First, in AE, the energy signal originates from the sample itself making its own signal, in response to stress. Second, the AE can detect the dynamic process because of its capability to detect movement or strain, whereas most of the other methods can detect existing geometrical discontinuities or fractures [20]. Thus, AE techniques have been applied to detect the crack location [21, 22] to quantify the degree of damage [23] and to determine the crack classification [24] in concrete structures.

These efforts, which will greatly improve our understanding of how concrete members reinforced with GFRP bars should be analysed, as well as the combination of these techniques, is expected to overcome the shortcomings of the respective techniques, increasing the efficiency of structural inspection and allowing for more frequent monitoring of structures.

2. AE monitoring technique

AE has been recently recognised as one of the most reliable passive tools for in-situ health monitoring of civil engineering RC structures [19]. It employs surface-mounted AE-sensors to capture the energy bursts in the form of the transient elastic-stress waves. These elastic waves are generated due to the rapid release of energy during deformation or crack propagation in RC structures during any type of loading [20, 25]. These AE sensors convert transient elastic waves into electrical signals. In AET, various AE parameters are extracted and used to correlate to damage initiation and progression in various kinds of infrastructures as well as localise and quantify it. Some of the key AE parameters reported for damage analysis are cumulative AE hits and AE counts, AE energy (MARSE), and AE signal strength are shown in **Figure 1**. Acoustic emission hits in AE bursts are described as the number of times an AE transient signal crosses the threshold value of the anticipated signals in a structure. As the cumulative AE hits and counts increase, it points towards damage progression in the structure and gives information about the intensity of the AE event. AE energy is the transient elastic energy released during an AE event and is measured as the area under the AE signal. A significant jump in the AE signal energy in the form of a 'Knee" indicates severe damage in the form of macro-cracking in the structure [21, 22].

Recent studies have also demonstrated the potential of AE techniques to detect the onset and propagation of damage/cracking in RC structures under flexural loading. Characterisation of cracks in plain and reinforced concrete beams subjected to flexural loading up to failure has also been reported [25]. It has been reported that as the level of damage in the RC beam increases, an increase in AE parameters of AE hits, counts, AE energy, rise time, and duration has been observed. AE parameters of average frequency (AF) and rise angle (RA) have been correlated with the cracking pattern and its type-tensile or shear cracks [27]. It has been observed that AE-based Ib-value along with RA and AF has been successfully used for the

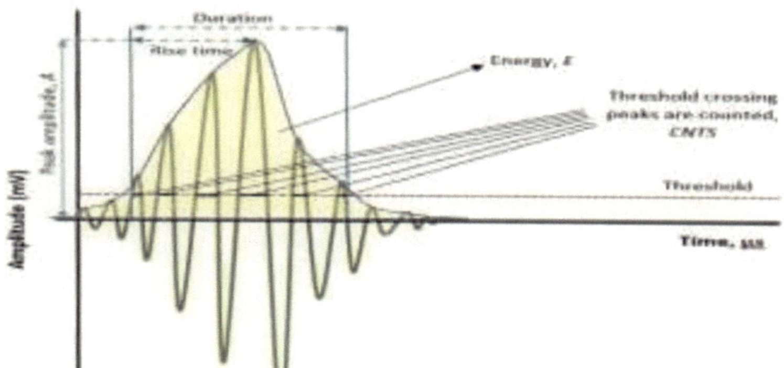

Figure 1.
Waveform of AE signal and its various parameter [26].

evaluation of flexural deformation of RC beams under cyclic loading [27]. AET has also been recently used for monitoring the fracture behaviour of different types of composite concrete beams [28]. AE not only determines accurately the onset of cracking and monitors the development of fracture but also indicates various kinds of damage and fracture modes in the form of de-bonding and concrete cracking in these beams. Hence, it can be concluded that AET has been established as a potential NDT tool for monitoring the performance of RC structures under loading when subjected to various types of damages. In this work, the efficacy of AET to understand and compare the failure pattern of steel and GFRP reinforced concrete is explored to establish its effectiveness as a potential NDT tool for concrete structures.

In this study, steel-RC and GFRP-RC beams with varying tension reinforcement were prepared and tested under the four-point bending test associated with AE equipment. The main objectives of this chapter are to examine the behaviour of cracks at each stage of the mechanical behaviour of the RC beams from loading to failure using the AE parameter analysis-based method. Moreover, this chapter also attempts to examine the effect of the changes in the varying percentage tension reinforcement ratio of steel-RC and GFRP-RC beam and level of damage on the parameters (cumulative AE hits, amplitude, rise angle, and A-FRQ) of the AE parameter analysis-based method. Furthermore, this chapter also aims to classify the crack types and classification of damage level occurs in two differently reinforced concrete beams along with varying percentage tension reinforcement ratio.

3. Experimental program and methodology

3.1 Specimen details and test matrix

RC beams having an effective length of 2000 mm with an overhang of 50 mm on each side with 150 mm × 230 mm cross-sectional dimensions were cast (**Figure 2**) using design mix proportions of 1:1.47:2.54 of cement, sand, and coarse aggregates with water-cement ratio of 0.46 using IS code method [30, 31]. The average compressive strength of the concrete used in both steel and GFRP reinforced concrete beams was experimentally obtained as 35.9 MPa. Moreover, mechanical properties of steel [32] and GFRP bars [33] were determined experimentally using Universal Testing Machine (Hung Ta Make, Taiwan with 1000 kN capacity) as shown in **Table 1**.

In the present study, two sets of beams were cast-one reinforced with traditional steel bars of Fe-500 grade bars denoted as (S-series) and the other reinforced with GFRP bars denoted as (G-series). It is important to note that S-series beams had both longitudinal as well as transverse reinforcement made of Fe 500 steel whereas

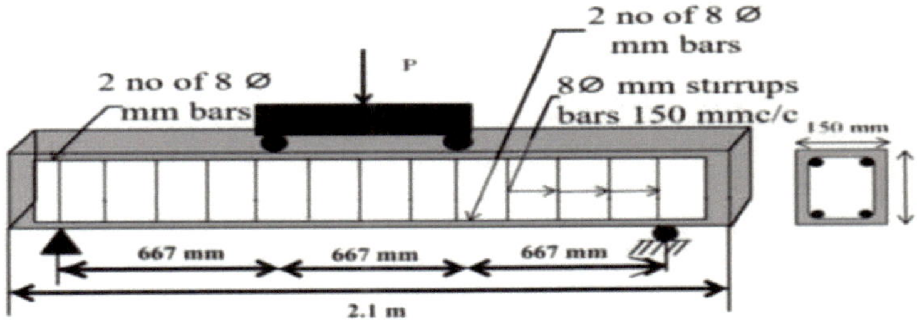

Figure 2.
Longitudinal and X-section details [29].

Bar diameter	Ultimate tensile strength (MPa)		Elastic modulus (GPa)		Ultimate strain	
(mm)	Steel	GFRP	Steel	GFRP	Steel	GFRP
8	530	866	201	41.3	0.0049	0.021
10	544	1092	205	43.7	0.0048	0.025
12	566	1219	210	41.20	0.0048	0.029

Table 1.
Mechanical properties of the reinforcing bars.

Series Code		ρ (%)	Asc	Ast
S-0.33-1 S-0.33-2	G-0.33-1 G-0.33-2	0.33%	2–8 mm Ø	2–8 mm Ø
S-0.33-3	G-0.33-3			
S-0.52-1 S-0.52-2 S-0.52-3	G-0.52-1 G-0.52-2 G-0.52-3	0.52%	2–8 mm Ø	2–10 mm Ø
S-1.11-1 S-1.11-2 S-1.11–3	G-1.11-1 G-1.11-2 G-1.11–3	1.11%	2–8 mm Ø	2–12 mm Ø

Asc = area of compression reinforcement; Ast = area of tension reinforcement.

Table 2.
Reinforcement details in the steel-RC and GFRP-RC beams.

G-series beams had both longitudinal as-well-as transverse reinforcement entirely made of GFRP bars.

The design of RC beams is based on the [8]. Steel-RC beams were designed as under reinforced [34] whereas GFRP-RC beams were designed as over reinforced [8]. The reinforcement ratio ($\rho = 100A_{st}/bd\%$) for each set of beams was varied as 0.33, 0.52, and 1.1% based on volumetric calculations. The steel-RC and GFRP-RC beams were identified according to the series. The arrangement is in the form of A-B-C, where A is the steel or GFRP-RC beam type, B is the steel or GFRP reinforcement ratio, and C is the name of the specimen which is denoted as numeric numbers 1, 2, 3. The reinforcement details of both S- and G-series are shown in **Table 2**. Three specimens of each beam series were cast to ensure repeatability of results but only one beam per type of each reinforcement ratio is explained in this research effort.

The experimental investigation involves testing of steel and GFRP reinforced concrete beams in four-point flexural loading which was displacement controlled at a rate of 0.01 mm/s (**Figure 3**). The loads were applied at L/3 from both supports using a steel spherical roller with a hydraulically controlled load cell (**Figure 3**). Mid-span deflections were measured using a Linear variable differential transformer (LVDT) attached underside of the RC beam and the load-deflection data was recorded by a high, speed data acquisition system. Before the actual AE monitoring, the AE sensors were checked for sensitivity using the pencil lead break test (PLB). After a successful PLB test, the wave velocity of concrete was set to 3.5×10^6 mm/s. To acquire AE signals, a threshold of 45 dB was set initially with a preamplifier gain of 40 dB as input. AE-win software was used to acquire the signals originating due to bending and subsequent cracking. The mechanical performance of the steel and GFRP reinforced beams was compared by studying load-deflection characteristics, failure modes, and the progression of visible cracking patterns and moment carrying capacity. For AE monitoring of the steel and GFRP-RC beams, six AE sensors (R6α, PAC Make) with a resonant frequency of 60 kHz were attached to the front (3 Nos) and the back-face surface of the beam (3 Nos) as shown in **Figure 3**. The AE sensors were attached to the beams using a Vaseline gel and held

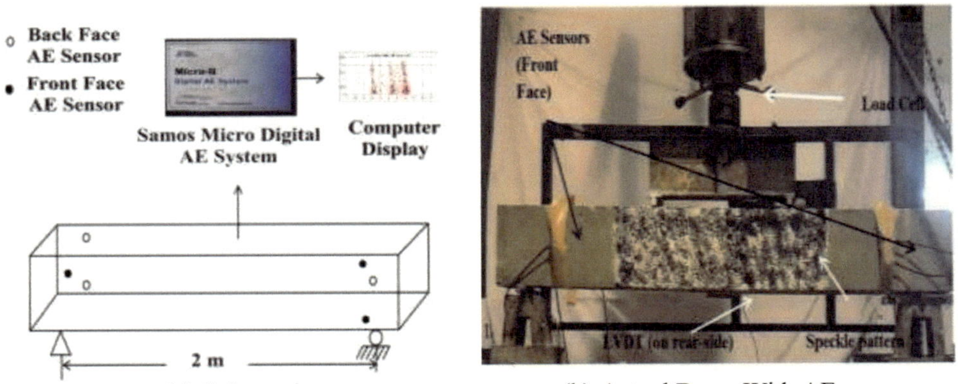

(a) Schematic (b) Actual Beam With AE sensor

Figure 3.
Acoustic emission monitoring setup [29]. (a) Schematic. (b) Actual beam with AE sensor.

in position using cello tape till the end of the experiment of steel and GFRP-RC beams. AE signals were recorded continuously during the entire duration of the loading of the beam. From the recorded AE signals, various AE waveform parameters of amplitude and number of AE hits, their expanse, and spread obtained using AE X-Y event plots have been used to study the variation in fracture and failure pattern of steel-RC and GFRP-RC beams. The speckle pattern shown in **Figure 3**, is used for digital image correlation (DIC) analysis which is the future scope of the work.

4. Results and discussions

4.1 Flexural performance of steel reinforced beams

The load-deflection plot of steel-RC beams is broadly classified into three regions un-cracked elastic, cracked-elastic, and plastic zones (**Figures 4** and 5). Initially, the applied loads as well as deflection are small and follow a linear relationship.

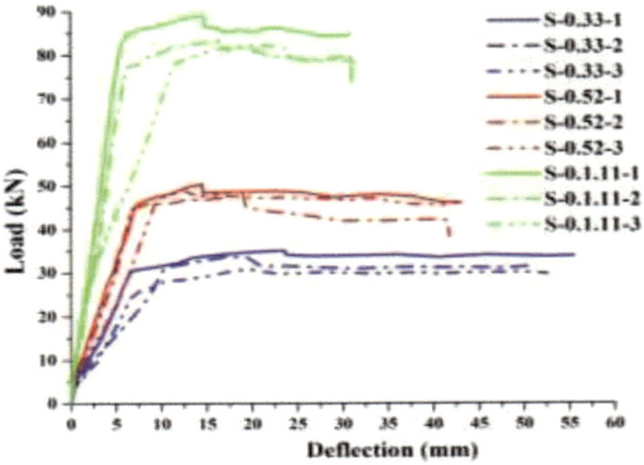

Figure 4.
Load v/s deflection plots for S-series beams.

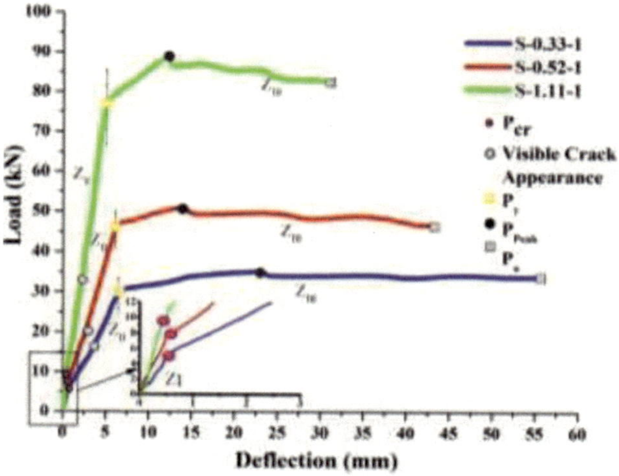

Figure 5.
Load v/s deflection plot for G-series beams.

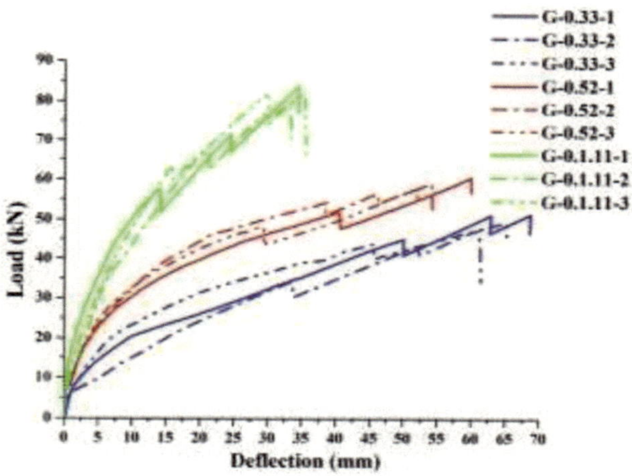

Figure 6.
Load v/s deflection plots for S-series beams (one beam per type).

This zone I is named uncracked elastic zone as shown in **Figure 6**. With further increase in loading, a significant change and reduction in stiffness of the beam are observed with the development of hairline cracks at cracking load (P_{cr}) of (5.58, 7.59, and 9.54) kN with a deflection (δ_{cr}) of (0.61, 0.60, and 0.43) mm for S-0.33-1, S-0.52-1, and S-1.11-1 beam, respectively (**Table 3**). These cracks progress along the sides of the beam at constant stiffness. The cracks initiate and start becoming visible at a load of (16, 20, and 30) kN for S-0.33-1, S-0.52-1, and S-1.11-1 beam, respectively, in the tensile zone of the beam. With further increase in loading, the cracks start propagating and appear in the form of distributed flexural and shear cracks leading to steel yielding at a load (P_y) of (28.93, 44.18, and 75.09) kN, with a deflection (δ_y) of (6.27, 6.85, and 4.94) mm in S-0.33-1, S-0.52-1, and S-1.11-1 RC beams, respectively as shown in **Table 3**. This part of the load-deflection plot from P_{cr} to P_y is termed as cracked-elastic zone II.

In zone III named plastic zone, the concrete section is cracked and ineffective in resisting the loads and the entire load is taken by steel and yields. It is marked by an increase in the mid-span deflection 23.22, 14.47, and 12.48 mm with a minor

Parameters	P_{cr} (kN)	δcr (mm)	P_y (kN)	δ_y (mm)	Ppeak (kN)	P_{ult} (kN)	δult (mm)	δA ult (mm)
S-0.33-1	**5.58**	**0.61**	**28.93**	**6.27**	**35.18**	**33.43**	**55.51**	53.13
S-0.33-2	4.41	0.54	30.84	10.23	33.70	31.58	51.23	
S-0.33-3	5.38	0.99	24.97	6.82	30.71	29.82	52.65	
G-0.33-1	**7.89**	**1.37**	**33.6**	**30.33**	**51.32**	**48.94**	**68.68**	65.14
G-0.33-2	6.41	0.99	39.94	45.63	49.34	44.13	65.27	
G-0.33-3	6.22	0.97	33.4	32.49	47.68	34.13	61.48	
S-0.52-1	**6.99**	**0.55**	**45.85**	7.34	**50.43**	**46.10**	**43.10**	42.59
S-0.52-2	5.81	0.52	43.75	6.54	48.26	38.40	41.77	
S-0.52-3	6.14	0.53	46.26	9.53	48.80	46.08	42.90	
G-0.52-1	**8.01**	**0.91**	**51.8**	**40.47**	**57.73**	**56.74**	**60.10**	55.52
G-0.52-2	7.56	0.81	54.33	38.77	59.32	51.96	54.26	
G-0.52-3	8.77	0.75	48.19	29.37	57.60	53.69	52.20	
S-1.11-1	**9.54**	**0.43**	**81.44**	**5.50**	**88.94**	**84.85**	**30.9**	31.42
S-1.11-2	10.17	0.71	75.42	6.03	83.41	72.52	31.02	
S-1.11–3	9.75	0.34	75.87	11.04	82.81	78.01	32.36	
G-1.11-1	**10.31**	**0.34**	**57.46**	**14.06**	**83.71**	**76.98**	**34.47**	34.52
G-1.11-2	9.55	0.76	66.84	20.78	79.51	70.33	33.39	
G-1.11–3	12.9	1.03	63.70	15.95	81.81	64.57	35.71	

Table 3.
Comparison of flexural parameters of steel and GFRP reinforced beams.

increase in load up to a peak load (P_{Peak}) of (35.18, 50.80, and 88.96) kN, pointing towards larger strain at the level of steel and increase in curvature of the cracked section with an increase in the percentage of steel. Further due to the strain hardening of steel, the beams fail at an ultimate load (P_u) of (33.43, 44.96, and 82.75) kN with a deflection (δ_u) of (55.51, 43.1, and 30.9) mm in for S-0.33-1, S-0.52-1, and S-1.11-1 RC beams, respectively. In general, it is observed that with the increase in the reinforcement ratio, the ultimate load-carrying capacity increases by approximately 34% in S-0.52-1 and 42% in S-1.11-1 beams as compared to S-0.33-1 indicating higher load carrying capacity with an increase in tensile reinforcement. Another important observation is a significant increase in the area under the load-deflection plot with an increase in the reinforcement ratio. It is also observed that plastic zone III reduces drastically with an increase in steel. The failure takes place at the much lower strain in the S-1.11-1 RC beam. All S-series RC beams specimens failed by steel yielding and followed by concrete crushing.

4.2 Flexural performance of GFRP reinforced beams

The mechanical behaviour of GFRP-RC beams in flexure is visibly different from steel-RC beams (**Figure 7**). Broadly, the behaviour of GFRP-RC beams exhibits a bilinear load-deflection response up to the failure without any yielding or ductility as experienced by steel reinforced beams. Initially, the load-deflection curve is perfectly linear and this zone is un-cracked elastic zone I. A noticeable decrease in the stiffness of the beam is observed with the formation of hairline cracks at a load (P_{cr}) of (7.89,

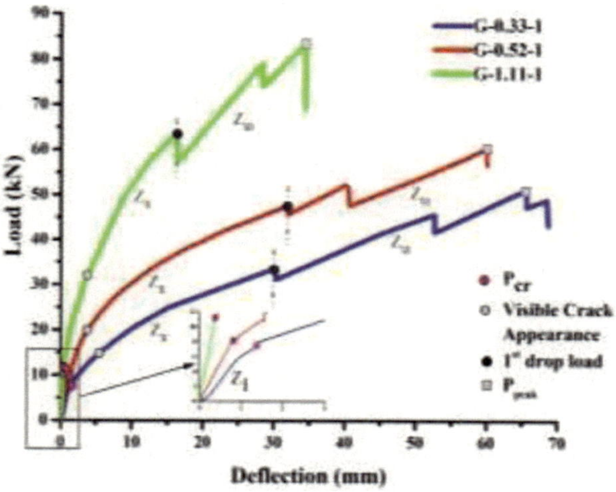

Figure 7.
Load v/s deflection plot for GFRP reinforced beams (one beam per type).

8.0 and 11.42) kN with a deflection (δ_{cr}) of (1.31, 0.9, and 0.39) mm in G-0.33-1, G-0.52-1, and G-1.11-1 RC beams, respectively. These minor cracks progress along the sides of the beam at constant stiffness. With an increase in the reinforcement ratio, initial bending stiffness also increases. As the load further increases, the cracks initiate and become visible at a load of (15, 20, and 33) kN for G-0.33-1, G-0.52-1, and G-1.11-1 RC beam in the pure bending region. Further with the increase in loading, cracks progress towards the compression zone, and this zone II is named cracked-elastic zone. The GFRP beams exhibit elastic response in this zone with the progression of flexural cracks. This trend continues till a first drop in the load-carrying capacity is observed at (33.6, 45.72, and 57.18) kN with a deflection of (30.33, 32.01, and 16.37) mm with increasing reinforcement ratio, pointing towards initiation of concrete crushing in G-0.33-1, G-0.52-1, and G-1.11-1 RC beams.

The failure is a typical flexural failure in the form of vertical flexural cracks in the pure bending zone along with their simultaneous spreading towards the entire length of the beam. Further the crushing of concrete progress with a sharp increase in the load-carrying capacity in zone III (concrete crushing zone). The beam continuous to carry load linearly with an increase in deflection until the second drop in load is observed at (41.7, 47.63, and 73.99) kN with a deflection of (52.68, 40.74, and 28.41) mm in G-0.33-1, G-0.52-1, and G-1.11-1 RC beams, respectively. The effective concrete section is highly reduced due to cracking and ineffective in resisting the tensile load and the beam fail at peak load of (P_{Peak}) (51.32, 60.47, and 83.71) kN for in G-0.33-1, G-0.52-1, and G-1.11-1 RC beam with the ultimate deflection (δ_u) (68.68, 60.09, and 34.47) mm. All the GFRP reinforced beams fail typically by concrete crushing since they are designed as over-reinforced beams to prevent the failure by GFRP rupture as expected in under-reinforced GFRP beams.

4.3 Effect of longitudinal tension reinforcement ratio on average mid-span deflection

In structural engineering, the term deflection is defined as the movement of a body from its original position under a force, load, or weight of the body itself. The effect of midspan deflection with the same concrete strength is intrinsically related to the longitudinal reinforcement ratio. As the longitudinal reinforcement ratio changes from 0.33, 0.52, and 1.1%, the mid-span deflection decrease in both steel and well as GFRP reinforced concrete beams. For example, in the case of steel-reinforced concrete beams, increasing the longitudinal reinforcement ratio changes from 0.33, 0.52, and 1.1% decreases the mid-span deflection by 53.15, 42.59, and 31.42 mm. On the other hand in the case of GFRP reinforced concrete beams decreases by 65.14, 55.52, and 34.43 mm. Hence, the overall comparison of steel and GFRP reinforced beams under static flexural loading is widely different. GFRP-RC beams exhibit higher deflections and lower crack widths in comparison to steel reinforced beams at the same reinforcement ratios (**Figure 8** and **Table 4**). The deflections are 22.55, 30.35, and 9.57% higher in G-0.33-1, G-0.33-1, and G-1.11-1 RC beams in comparison to S-0.33-1, S-0.52-1, and S-1.11-1 RC beams. This is because of the low elastic modulus of GFRP bars as compared to steel bars.

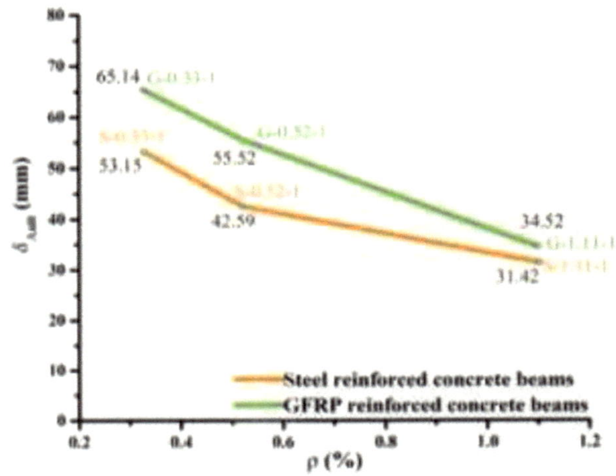

Figure 8.
Variation in average maximum mid-span deflection Vs 'ρ'.

RC specimen	Mexp (kN-m)	Mth (kN-m)	Mth/ Mexp	RC specimen	Mexp (kN-m)	Mth (kN-m)	Mth/ Mexp
S-0.33-1	11.73	7.83	0.66	G-0.33-1	17.11	15.33	0.93
S-0.33-2	11.23	7.83	0.70	G-0.33-2	16.45	15.33	0.93
S-0.33-3	10.24	7.83	0.76	G-0.33-3	15.90	15.33	0.96
Average	**11.06**	**7.83**	**0.70**	**Average**	**16.48**	**15.33**	**0.94**
S-0.52-1	16.09	12.63	0.78	G-0.52-1	19.25	18.83	0.97
S-0.52-2	16.27	12.63	0.77	G-0.52-2	19.78	18.83	0.95
S-0.52-3	16.81	12.63	0.75	G-0.52-3	19.20	18.83	0.98
Average	**16.39**	**12.63**	**0.76**	**Average**	**19.41**	**18.83**	**0.96**
S-1.11-1	27.81	23.95	0.86	G-1.11-1	27.91	25.63	0.95
S-1.11-2	29.66	23.95	0.80	G-1.11-2	26.51	25.63	0.96
S-1.11–3	27.61	23.95	0.86	G-1.11–3	27.28	25.63	0.93
Average	**28.36**	**23.95**	**0.84**	**Average**	**27.23**	**25.63**	**0.94**

Table 4.
Moment capacities of steel and GFRP reinforced beams.

4.4 Effect of longitudinal reinforcement ratio on experimental moment carrying capacities

The experimental *moment* carrying *capacity* is a maximum *bending moment* that can be resisted by a beam or any other structural member before it fails in *bending*. The effect of experimental moment carrying capacities with the same concrete strength is essentially related to the longitudinal reinforcement ratio.

As the longitudinal reinforcement ratio changes from 0.33, 0.52, and 1.1% the experimental *moment* carrying *capacity* increases in both steel and well as GFRP reinforced concrete beams. For example in the case of steel-reinforced concrete beams, increasing the longitudinal reinforcement ratio changes from 0.33, 0.52, and 1.1% increases the experimental *moment* carrying *capacity* by 11.06, 16.39, and 28.36 kN-m. On the other hand in the case of GFRP reinforced concrete beams increases by 16.48, 19.41, and 27.23 kN-m as shown in **Figure 9** and **Table 4.** Overall, it is observed that with the increase in the longitudinal reinforcement ratio, the ultimate load-carrying capacity increases by approximately 34% in G-0.33-1 and 42% in G-0.52-1 beams in comparison to S-0.33-1 and S-0.52-1. It is pointing towards higher tensile strength of GFRP bars in comparison to steel reinforced beams. But as the reinforcement ratio increases from 0.52 to 1.1% there is a slightly 5.09% decrease in load-carrying capacity in G-1.11-1 beam as compared to S-1.11-1 beam due to increase brittleness with higher ρ. It points towards a lower modulus of elasticity of GFRP bars as compared to steel bars.

4.5 Modes of failure

The design of both steel-RC and GFRP-RC is based on ACI Code [8, 34]. The GFRP-RC beams were designed as over-reinforced beams with a reinforcement ratio of 0.0052 [8] which was greater than the balanced reinforcement ratio of 0.00308 [34]. The steel-RC beams were designed as under-reinforced beams having a reinforcement ratio (0.0052) less than the balanced reinforcement ratio of 0.02. A compression failure for the GFRP-RC beams and a tension failure for the steel beams were expected during flexural testing. The observed modes of failure of

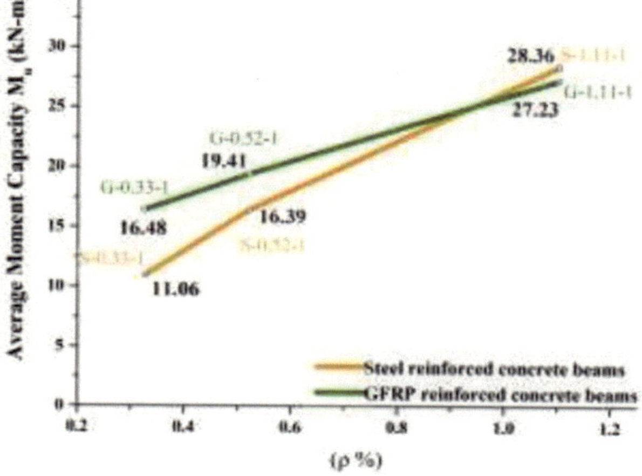

Figure 9.
Variation in an average experimental moment carrying capacities Vs 'ρ'.

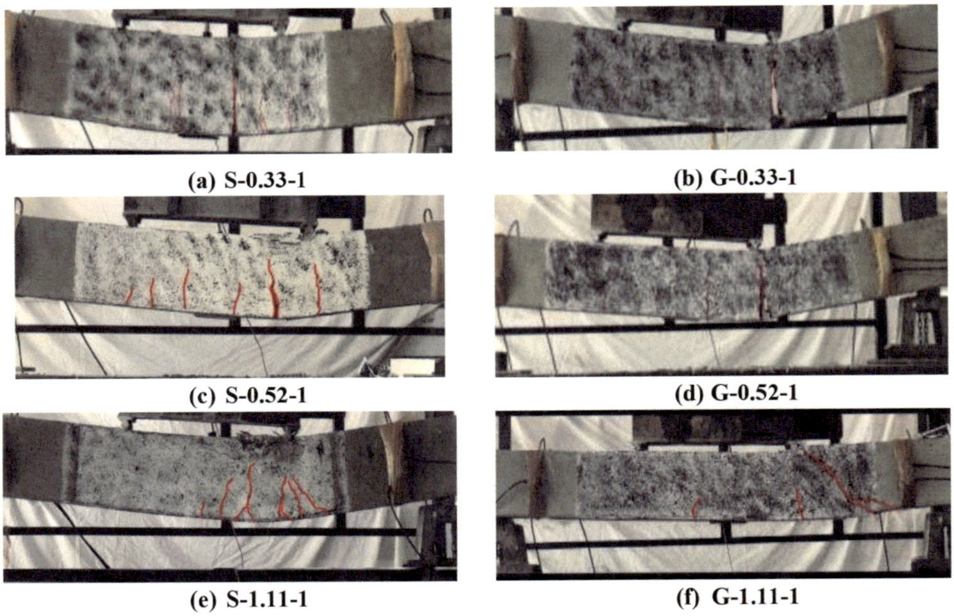

Figure 10.
Failure modes in steel-RC and GFRP-RC beams. (a) S-0.33-1. (b) G-0.33-1. (c) S-0.52-1. (d) G-0.52-1. (e) S-1.11-1. (f) G-1.11-1.

steel-RC and GFRP-RC beams are presented in **Figure 10.** Steel-RC beam failed by the crushing of concrete after the tension reinforcement yielded (**Figure 10a**, **c**, and **e**) whereas the GFRP-RC beam failed typically in shear followed by concrete crushing **(Figure 10b**, **d**, and **f**) since they are designed as over-reinforced beams to prevent their failure by GFRP rupture. This indicates that even though both steel-RC and GFRP-RC beams have the same area of tension reinforcement (A_{st}) ratio, GFRP-RC beams experience a different mode of failure as compared to the steel-reinforced concrete beam. Therefore, some modification in the design has to be considered when the GFRP bar is to be used as reinforcement.

A comparison of theoretical and experimental moment carrying capacities of steel-RC and GFRP-RC beams is presented in **Table 4.** The ratio of M_{th}/M_{exp} is less than 1 for both steel-RC and GFRP-RC beams. Thus, for design purposes, the strength reduction factor (Ø) for all over reinforced GFRP reinforced beams is calculated by Eq. (1). The theoretical moment of resistance (M_{th}) in the case of the GFRP-RC beam is calculated by Eq. (1) [8].

$$M_{th} = \rho_f f_f \left[1 - 0.59\rho_f \frac{f_f}{f'c}\right] bd^2 \quad (1)$$

On the other hand, the theoretical moment of resistance (M_{th}) in the case of the steel-RC beam is calculated by the formula given in Eq. (2) [34]

$$M_{th} = \rho_f f_f \left[1 - 0.59\rho_f \frac{f_y}{f'c}\right] bd^2 \quad (2)$$

where M_{th} = theoretical bending moment resistance, ρ_f = reinforcement ratio, b = width of the beam, d = effective depth of the beam, f'_c = Design characteristic concrete compressive strength. Where, ρ_{fb} = balanced reinforcement ratio, ρ_f = actual reinforcement ratio and $\emptyset$ = 0.65 for $\rho_f \geq 1.4\ \rho_{fb}$ [8]. For under-reinforced steel beams, the strength reduction factor is taken as 0.9 (ACI 319, (2019)). The fracture/failure and cracking pattern in the two differently reinforced beams along with a varying percentage of tension reinforcement ratio are further studied using acoustic emission in Section 5 respectively.

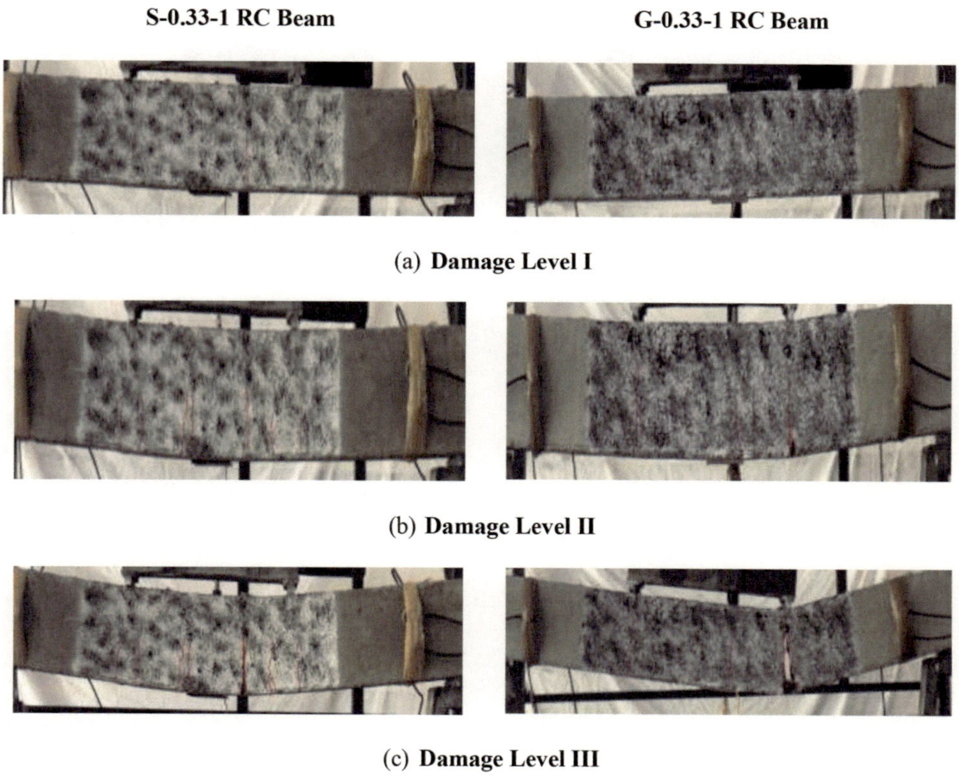

Figure 11.
Typical crack patterns. (a) Damage level I. (b) Damage level II. (c) Damage level III.

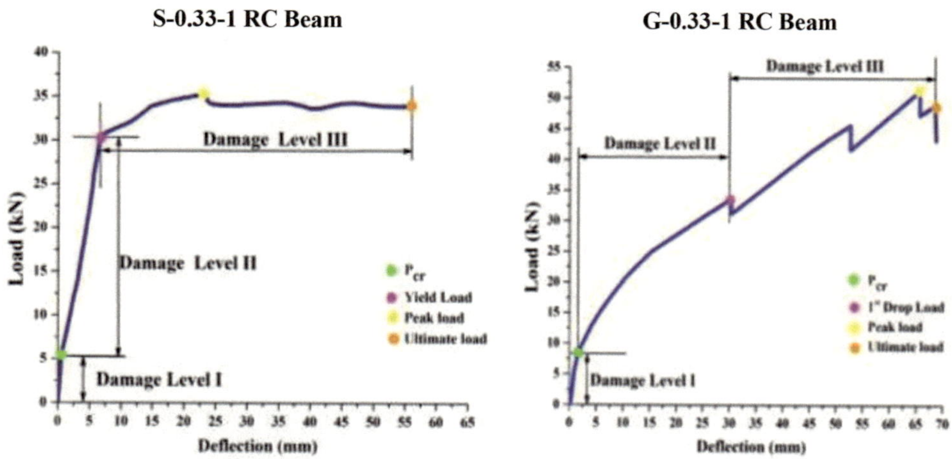

Figure 12.
Damage level classification.

4.6 Damage classification

From the load-deflection plots of steel-RC and GFRP-RC beams, as shown in **Figures 12, 14**, and **16** the development of cracking patterns as shown in **Figures 11, 13**, and **15**) in the two beams can be classified into three damage levels. Damage level I refers to the phase when the invisible cracking occurs. This damage zone is differentiated from the un-damaged state by the formation of

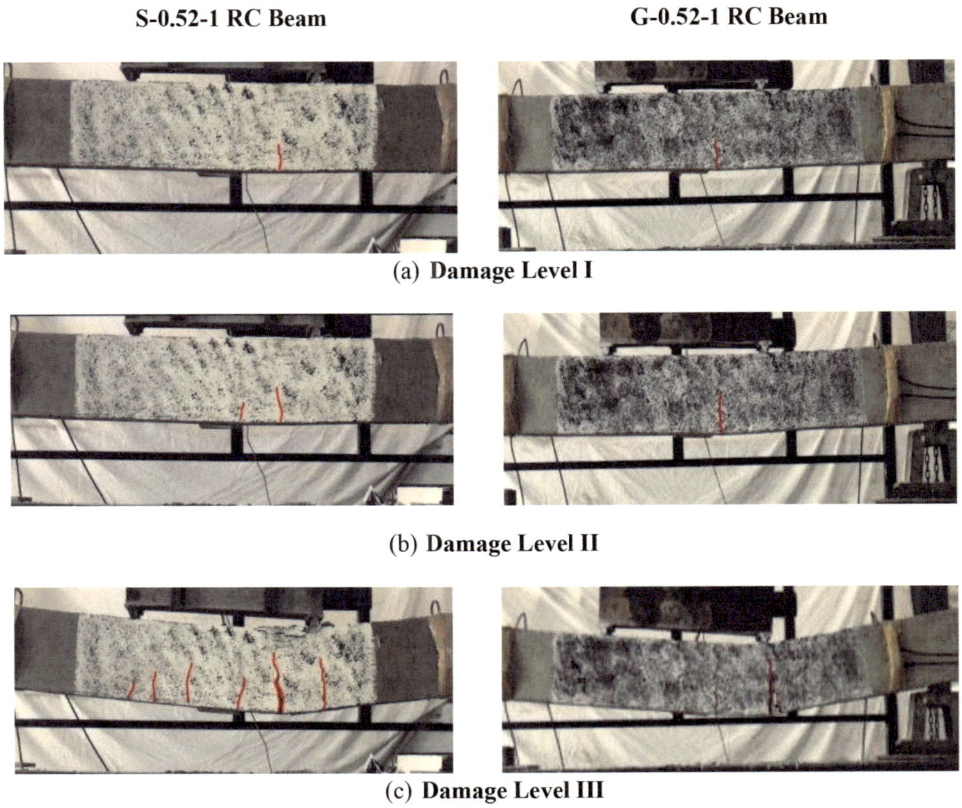

Figure 13.
Typical crack patterns. (a) Damage level I. (b) Damage level II. (c) Damage level III.

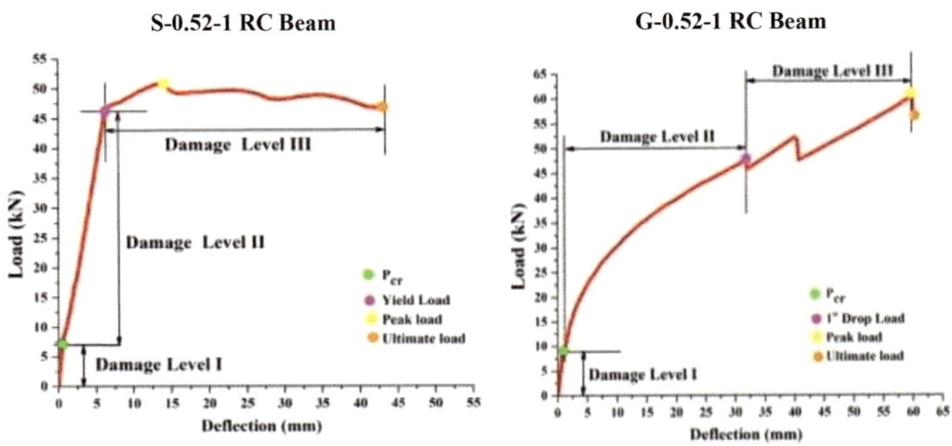

Figure 14.
Damage level classification.

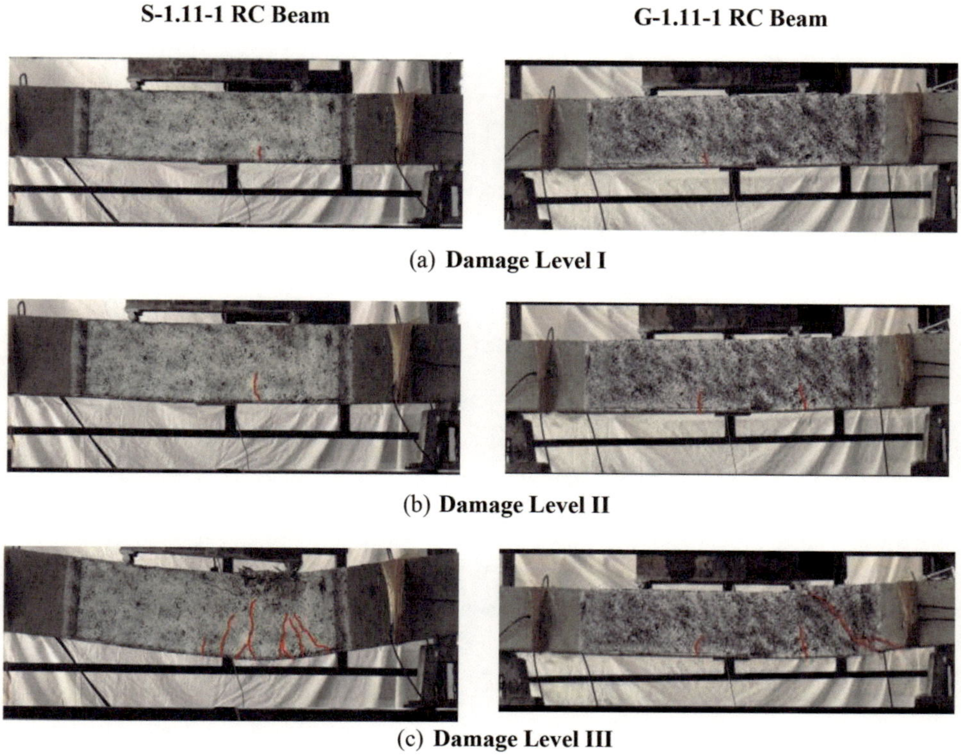

Figure 15.
Typical crack patterns. (a) Damage level I. (b) Damage level II. (c) Damage level III.

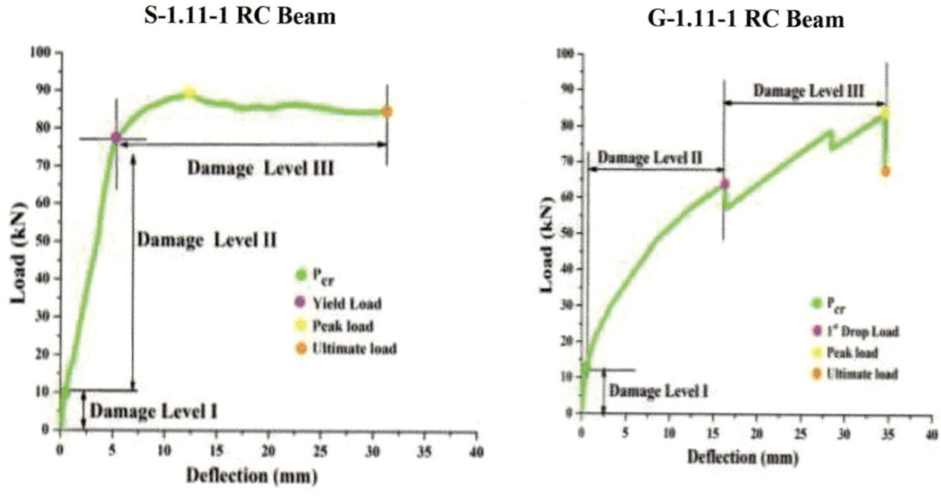

Figure 16.
Damage level classification.

visible hairline cracks and a distinct decrease in the stiffness of the beam is observed. *Damage level II* refers to the phase between the formation of hairline cracks and the stage of steel yielding in the case of the steel-RC beams and 1st drop-in load in case of GFRP-RC beam and further leading to the formation of distributed flexural and shear cracks. *Damage level III* refers to the phase between the steel yielding and final failure caused due to concrete crushing in the SB beam. In the case of GB beams, this refers to the phase between 1st drop in load and the final

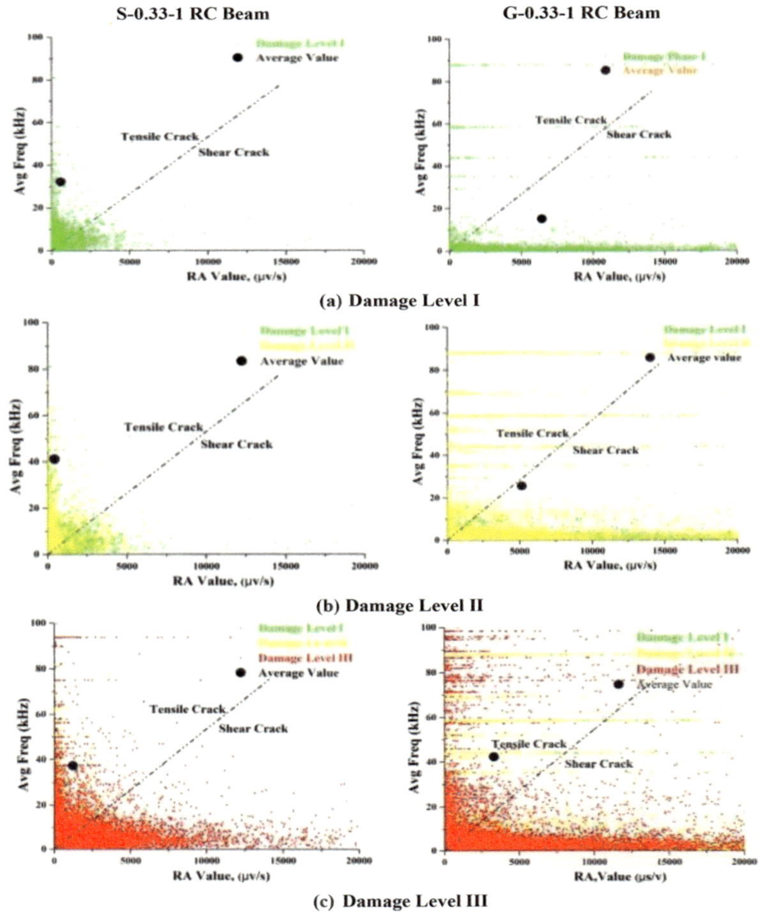

Figure 17.
Variation in AF Vs RA values at different levels of damages. (a) Damage level I. (b) Damage level II. (c) Damage level III.

failure due to the crushing of compressive concrete. The formation of crack patterns at different levels of loading and the development of the different damage levels in steel-RC and GFRP-RC beams are shown in **Figures 11–16** respectively.

5. Acoustic emission results

5.1 Average frequency (AF) and rise angle (RA)

A parametric analysis was performed between AF and RA values by moving averages based on over 100 AE hits [20, 25]. In the present case, the proportion of AF and RA values are likewise set to 1:200. The plot of Average frequency (AF) and rise angle (RA) values for all three damage levels, for S-series and G-series RC beams, is shown in **Figures 17–19** respectively. AF-RA value plot gives a fair indication of the cracking modes in steel-RC and GFRP-RC beams and is used for crack classification. The diagonal line represents the transition line between tensile and shear cracks and is used as a reference line for crack classification.

In damage level I, the plot of AF-RA for the S-0.33-1, S-0.52-1, and S-1.11-1 RC beam **(Figures 17(a), 18(a)**, and **19(a))** shows the development of cracks initially

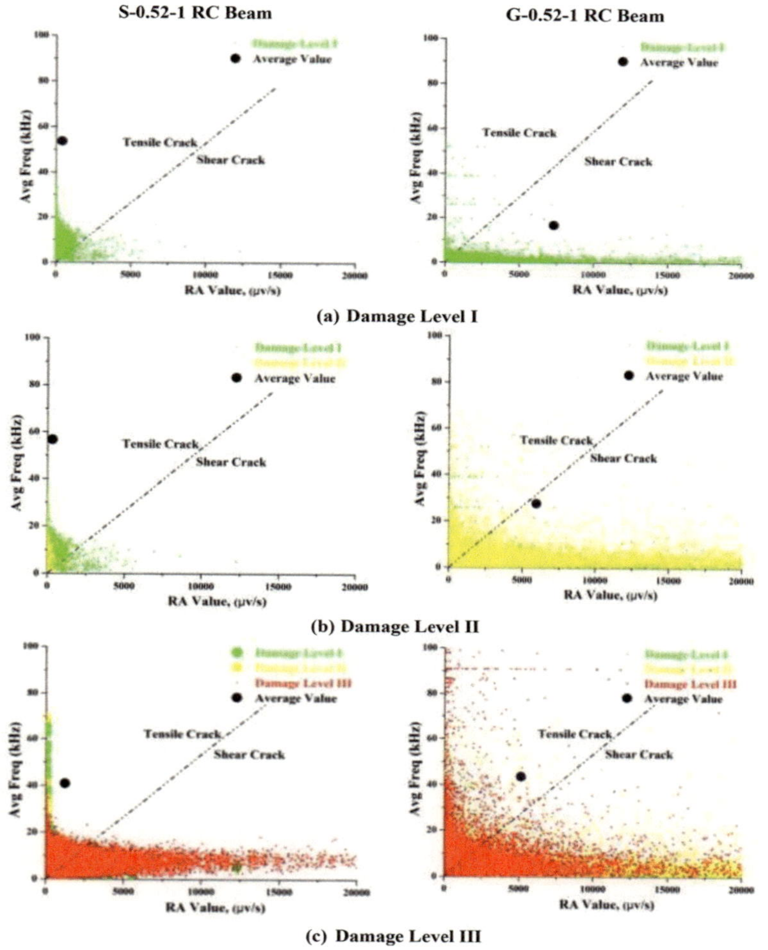

Figure 18.
Variation in AF vs. RA values at different levels of damages. (a) Damage level I. (b) Damage level II. (c) Damage level III.

due to tensile cracking whereas for the G-0.33-1, G-0.52-1, and G-1.11-1 RC beam, it is in shear cracking mode (**Figures 17(a)**, **18(a)**, and **19(a)**). A high average AF value of 32.2, 53.73, and 50.64 kHz with a lower RA value of 579.9, 398.78, and 205.0 µs values were noticed in the S-0.33-1, S-0.52-1, and S-1.11-1 RC beam as against lower average AF value of 15.0, 16.20, and 26.73 kHz and a higher RA value of 6432.5, 7324.80, and 10752.52 µs for the G-0.33-1, G-0.52-1, and G-1.11-1 RC beam (**Tables 5-7**), respectively. These values indicate that the steel-RC beams can resist and bridge the cracks better owing to the perfect bond between concrete and steel and the high modulus of elasticity of steel bars as compared to GFRP reinforced concrete beams. In this damage level, the hairline cracks were visible in both beams.

With the increasing load in damage level II, invisible cracking is observed in the steel-RC beam at higher AF and smaller RA value indicating tensile cracking mode whereas, in the case of the GFRP-RC beam, a reversed trend is observed with a slight increase in average AF value and drop in RA value pointing towards a shift from shear to tensile cracking in the GFRP-RC beam (**Figures 17(b)**, **18(b)**, and **19(b)**). Further, in damage level III, a slight decrease in the average AF value of 37.3, 40.99, and 47.41 kHz with a minute increase in the RA value of 1327.0, 1225.25, and 454.92 µs value was noticed in the S-0.33-1, S-0.52-1, and S-1.11-1 RC beam. A

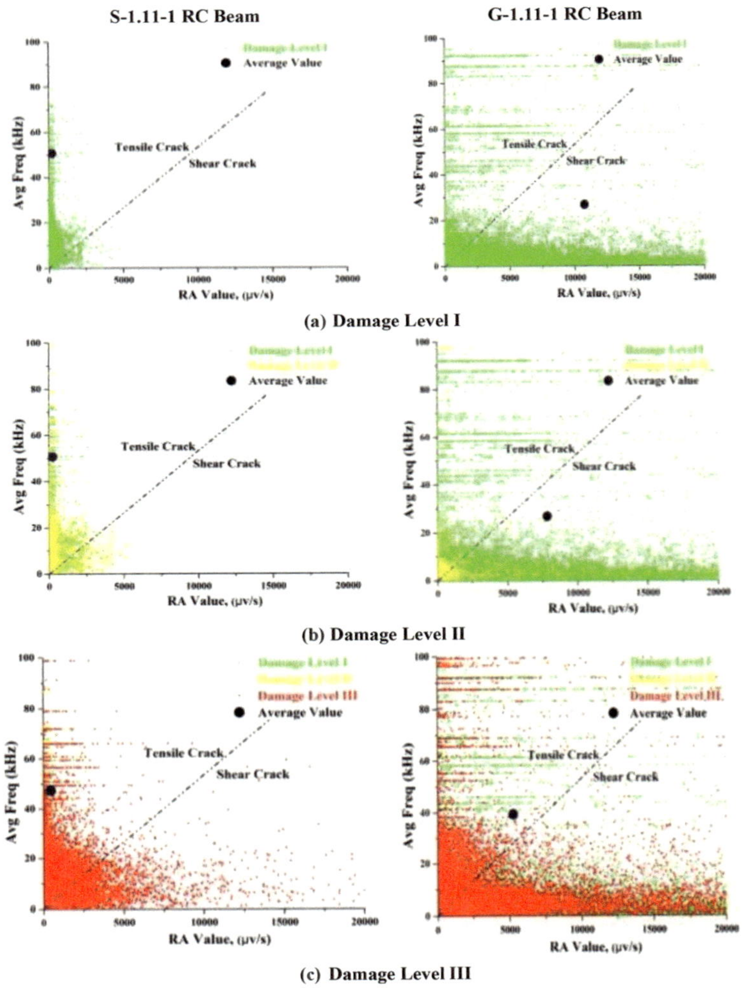

Figure 19.
Variation in AF vs. RA values at different levels of damages. (a) Damage level I. (b) Damage level II. (c) Damage level III.

Damage level	Rise angle		Average frequency	
	S1	G1	S1	G1
I	579.9	6432.5	32.2	15.0
II	427.9	5157.2	41.1	25.5
III	1327.0	3317.6	37.3	42.2

Table 5.
Variation in RA and AF values in S-0.33-1 and G-0.33-1 RC beams.

Damage level	Rise angle		Average frequency	
	S2	G2	S2	G2
I	398.78	7324.8	53.73	16.20
II	315.21	5956.17	56.72	27.65
III	1225.257	5130.68	40.99	43.66

Table 6.
Variation in RA and AF values in S-0.52-1 and G-0.52-1 RC beams.

Damage level	Rise angle		Average frequency	
	S3	G3	S3	G3
I	205	10752.52	50.64	26.73
II	238.79	7829.34	50.71	26.81
III	454.922	5255	47.41	39.4

Table 7.
Variation in RA and AF values in S-0.1.11-1 and G-1.11-1 RC beams.

plausible explanation for this can be attributed to the reduction in the cross-sectional area due to the yielding of steel bars in the steel-RC beam.

On the other hand, a continuous increase in AF value of 42.2, 43.66, and 39.40 kHz with a significantly lower RA value of 3317.6, 5130.68, and 5255 μs was noticed in the G-0.33-1, G-0.52-1, and G-1.11-1 RC beam (**Figures 17(c), 18(c)** and **19(c)**). AFRA plots at the failure point suggest that the steel-RC beam experiences flexural cracks localisation as shown in **Figure 10a, c**, and **e**, whereas the GFRP-RC beam experiences shear cracking as also observed visually as shown in **Figure 10b, d**, and **f.**

Hence, AF-RA plots can be exploited to predict the initiation and progression of invisible and visible crack formation in concrete as indicated by the density of dots in the plot of the AFRA value. Hence, the AE plot of AF and RA value can effectively demonstrate the variation in initiation and progression of damage, classification of cracking, and failure modes in steel as well as GFRP-reinforced beams.

5.2 AE XY plots monitoring

To further understand the sequences of AE events, their locations within the steel-RC and GFRP-RC beam have been plotted at different times in the form of event maps. The locations of the AE events during the entire process of damage are presented in X-Y plots and these AE XY-event plot profiles were obtained by using AE-win software (**Figures 20–28**). It may be recalled that the cracks can be located only in the zones covered by the AE sensors. Every crack is labelled as an event recorded by three or more sensors. The red spawn (dots) in AE plots represents the location of each AE event recorded to indicate the frontal surface condition of the steel-RC and GFRP-RC beams at different stages of damage under flexural loading. AET is supposed to increase the efficiency of structural inspection by indicating the initiation and progression of cracking and the surface strains in the two types of beams respectively.

It is apparent from **Figures 20(a), 23(a)**, and **26(b)** that, during the damage level I i.e., at cracking loads of ~5.58, 6.99, and 9.54 kN for S-0.33-1, S-0.52-1, and S-1.11-1 RC beam and ~7.89, 8.01, and 10.31 kN for G-0.33-1, G-0.52-1, and G-1.11-1 RC beam respectively. It is indicated by the appearance of AE events in the XY-plot at the same instant which suggests the formation of invisible cracks in steel-RC and GFRP-RC at the same location as shown in **Figures 20(b), 23(b)**, and **26(b)**. Hence, it can be concluded that invisible cracking which is not visible to the naked eye can be reliably displayed by AE XY- event plots.

Further, with an increase in loading, it is visually observed that at a yield load (P_y) of ~28.93, 45.85, and 81.44 kN in S-0.33-1, S-0.52-1, and S-1.11-1 RC beam the earlier invisible crack starts to become visible and eventually coalesce together to form visible cracks propagating vertically upwards. This indicates the progression of damage to level II as shown in **Figures 21(a), 24(a)**, and **27(a)**. These cracks do

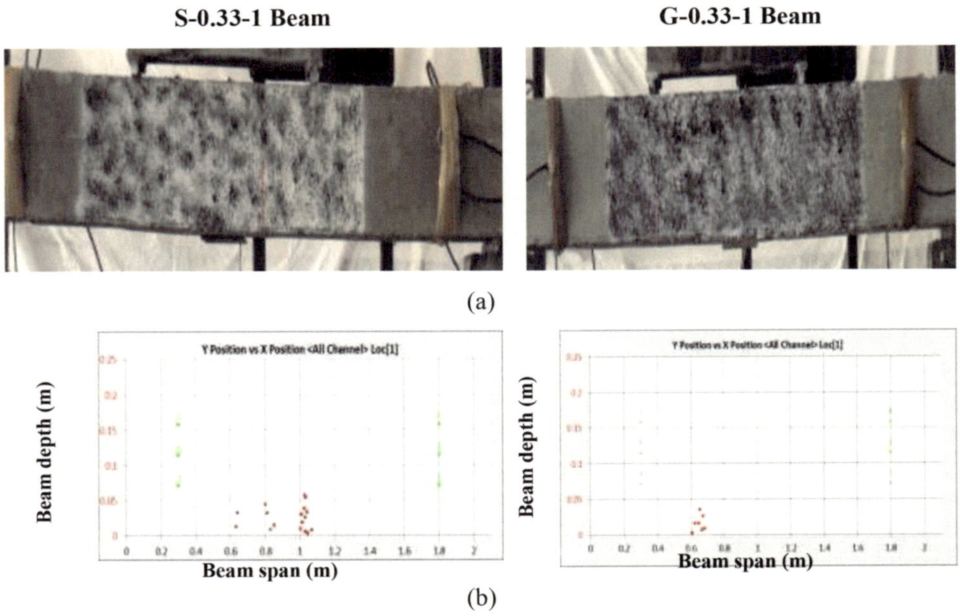

Figure 20.
(a) Beam sample (b) AE XY plots at damage level I.

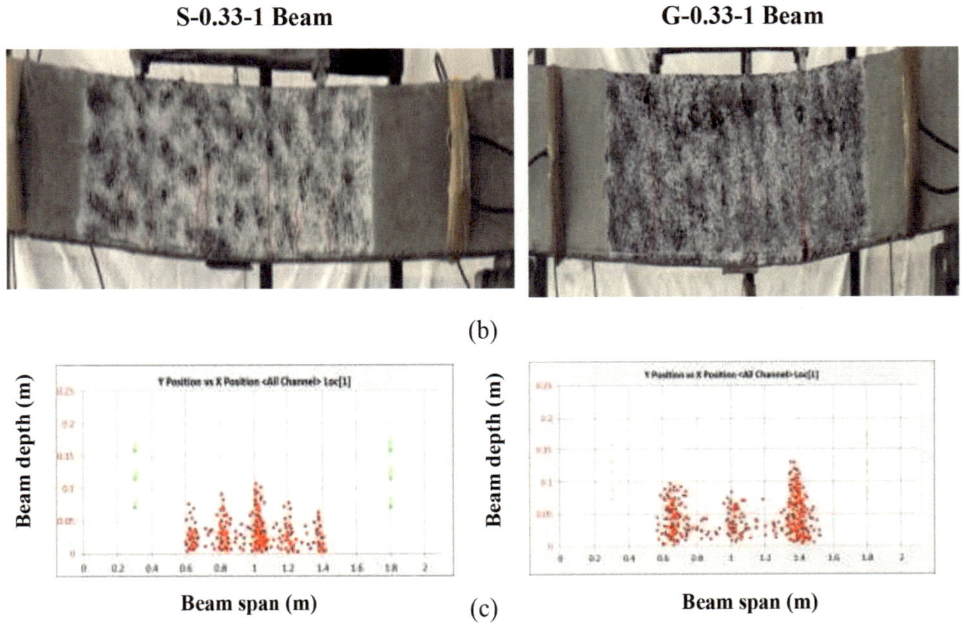

Figure 21.
(a) Beam sample (b) AE XY plots at damage level II.

not result in sudden failure of the beam as their propagation is arrested by the presence of shear stirrups. On the contrary, in the case of the GFRP-RC beams, owing to the elastic behaviour of GFRP bars, the beam continues to carry load linearly and invisible cracking is observed at a load of ~33.6, 51.80, and 57.46 kN. On the other hand, the AE event plot shows the congregation of red dots pointing towards the coalescence of invisible cracks into visible cracks at approximately (1.2, 1.2, and 1.1) m and (1.3, 1.2, and 1.1) m distance from the left support for S-0.33-1,

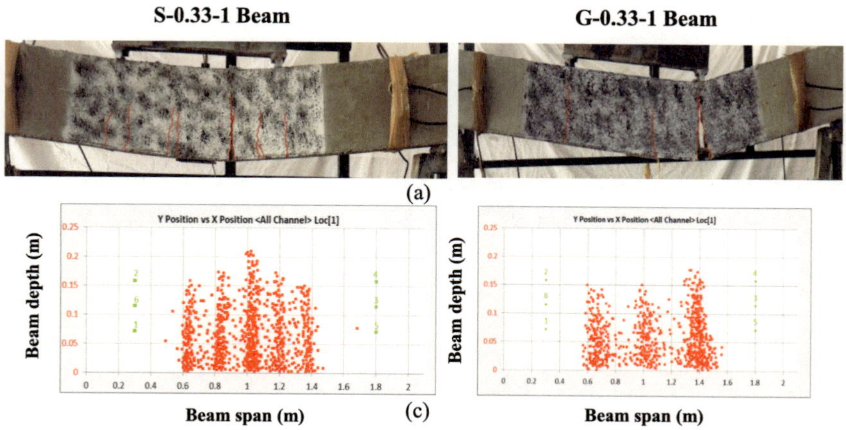

Figure 22.
(a) Beam sample (b) AE location- XY plots at damage level III.

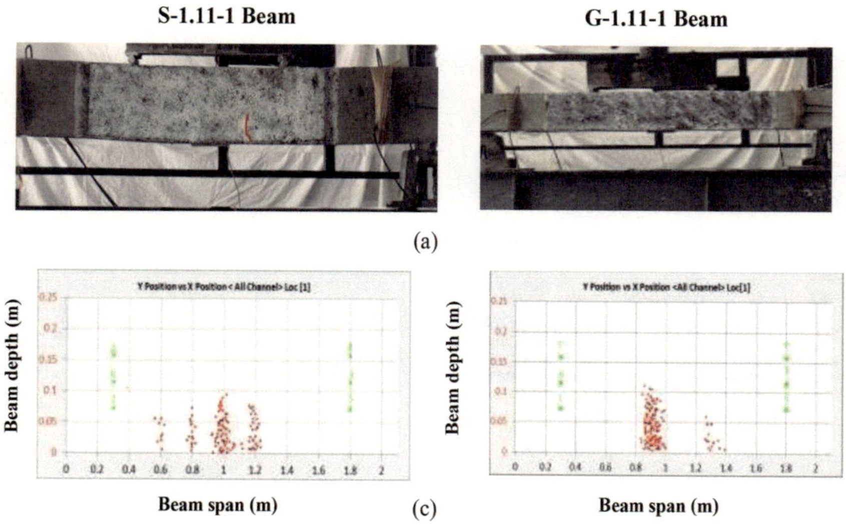

Figure 23.
(a) Beam sample (b) AE XY plots at damage level I. (c) Beam span (m).

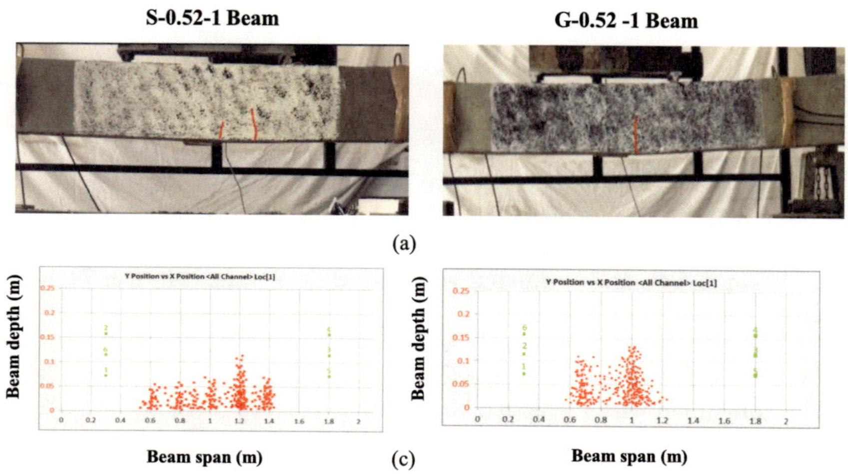

Figure 24.
(a) Beam sample (b) AE XY plots at damage level II.

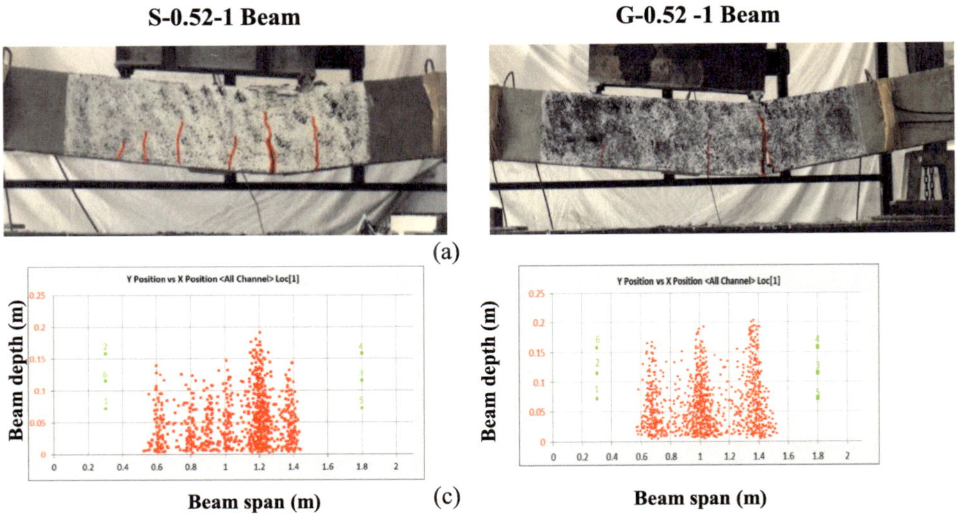

Figure 25.
(a) Beam sample (b) AE location- XY plots at damage level III.

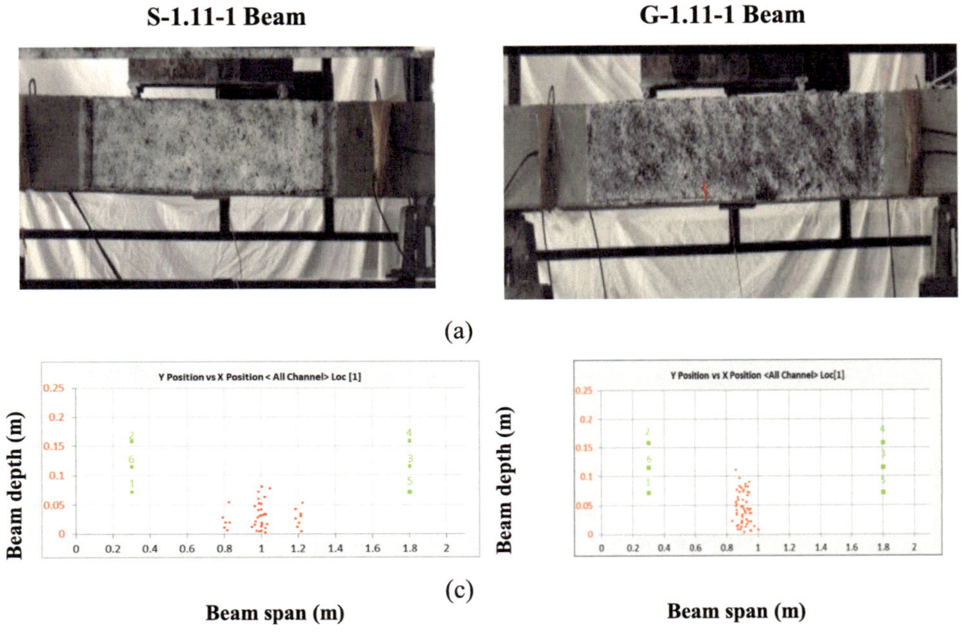

Figure 26.
(a) Beam sample (c) AE XY plots at damage level I.

S-0.52-1, and S-1.11-1 RC and G-0.33-1, G-0.52-1, and G-1.11-1 RC beams. This is in a close match with the actually cracked beam and the same can also be observed in the actual beam sample and compared with the AE X-Y plots (**Figures 21(b)**, **24(b)**, and **27(b)**).

With further loading i.e. damage level III, the steel bars yield leading to flexural failure followed by concrete crushing (**Figures 22(a)**, **25(a)**, and **28(a)**). The same can also be confirmed with the actual beam sample. Moreover, the G-0.33-1, G-0.52-1, and G-1.11-1beams fail typically in shear followed by concrete crushing at damage level III. The corresponding ultimate load of ~48.94, 56.74, and 76.98 kN in G-0.33-1, G-0.52-1, and G-1.11-1beams and is also depicted by extremely dense AE event plots at the same locations at 1.33, 1.33, and 1.33 m from the left support in

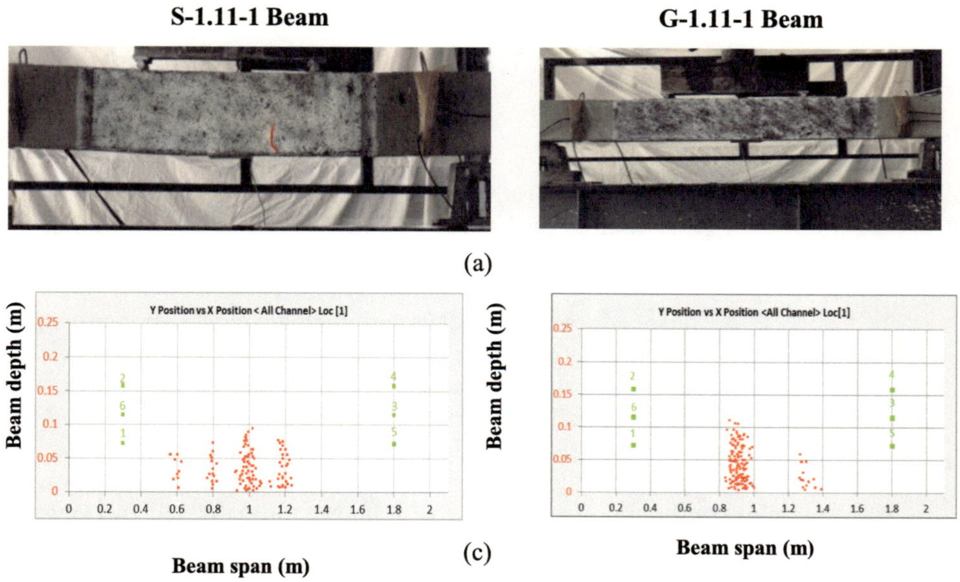

Figure 27.
(a) Beam sample (c) AE XY plots at damage level II.

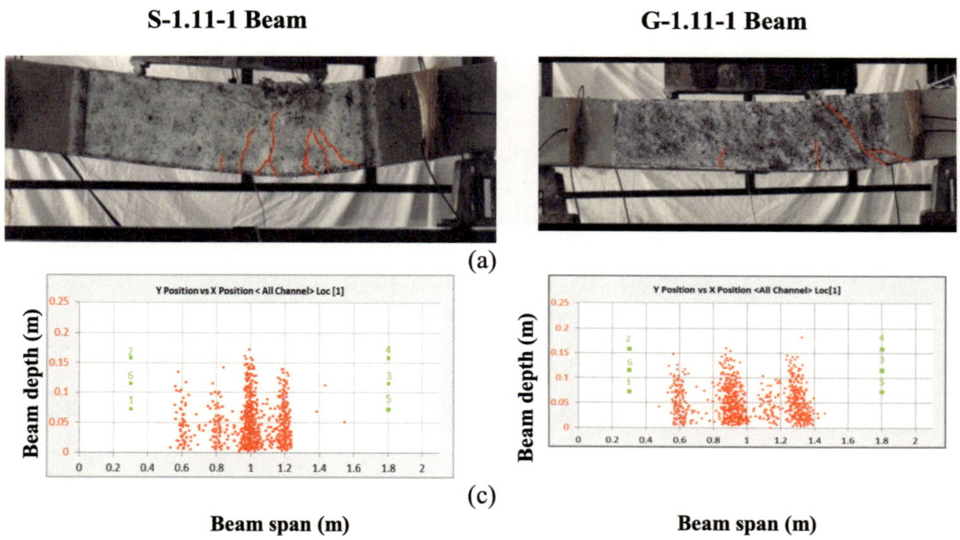

Figure 28.
(a) Beam sample (b) AE location- XY plots at damage level III.

Figures 22(b), **25(b)**, and **28(b)**. A clear trajectory of transverse vertical cracks is also observed from the AE event plot. This is also validated by the image of the actually cracked beams. Thus, AE events maps provide a reliable and real-time indication of the initiation and progression of cracking inside concrete in steel as well as GFRP-RC beams undergoing flexural loading.

Hence, it can be concluded that AE has the potential to serve as an online non-destructive monitor that can map the progression of cracking in RC structures. Various AE parameters like cumulative AE hits and their amplitudes, the plot of AF, and RA can effectively capture the initiation and progression of cracking in the steel and GFRP-RC beams, much before these are visible to the naked eye. Moreover, AE XY-plot has ample potential to serve as an effective tool to monitor cracking in the terms of AE XY-plot, and much before the actual cracking is visible to the naked eye. Hence, the advanced AE tools can be effectively used in conjunction with NDE of various types of RC structures incorporating steel as well as GFRP bars.

6. Conclusion

The focus of this chapter is to critically investigate and differentiate between the flexural and cracking behaviour of steel-RC and GFRP-RC beams utilising a non-destructive technique of AE. The following conclusions are drawn from the study:

- The load-deflection plot of steel and GFRP reinforced beams shows contrasting profiles. With increasing reinforcement ratio, steel-reinforced beams show typically an increase in ultimate load-carrying capacity, shrinking plastic zone with reduced ductility, and failure taking place at the much lower strains. All the steel-RC beams failed in flexure followed by concrete crushing. On the other hand, GFRP reinforced beams exhibit bi-linear load-deflection response up to the failure without any yielding or with higher ultimate load-carrying capacities and higher deflections as compared to steel reinforced beams indicating enhanced ductility and ultimate strength. All the GFRP-RC beams failed in shear and were followed by concrete crushing.

- Surface-mounted passive AE sensors give an early warning of initiation of micro-cracking inside concrete well depicted by low amplitude AE hits with the progression of failure in the form of macro cracking displayed by AE hits of larger amplitudes in steel as well as GFRP reinforced concrete beams. The amplitude and the cumulative number of AE hits are well indicative of the initiation and progression of invisible and visible cracking in the RC beam in the form of well-defined AE activity phases. A striking contrast in the AE activity is observed in steel and GFRP reinforced beams.

- AE XY-event plots give a pictorial representation of various stages of onset, initiation, and progression of cracking in steel and GFRP reinforced concrete beams in terms of their origin, location their rates of growth. A close matching and coherence are observed between the AE XY-event map with the micro and macro-cracks visually observed in the actual beams at various stages of loading.

Author details

Gaurav Sharma[1*], Shruti Sharma[1] and Sandeep Kumar Sharma[2]

1 Civil Engineering Department, Thapar University, Patiala, India

2 Department of Mechanical Department, Thapar University, Patiala, India

*Address all correspondence to: sharma9988gaurav@gmail.com

References

[1] Valdez B, Ramirez J, Eliezer A, Schorr M, Ramos R, Salinas R. Corrosion assessment of infrastructure assets in coastal seas. Journal of Marine Engineering & Technology. 2016;**15**(3): 124-134

[2] Bhaskaran R, Bhalla L, Rahman A, Juneja S, Sonik U, Kaur S, et al. An analysis of the updated cost of corrosion in India. Materials Performance. 2014; **53**(8):56-65

[3] Gu L, Meng XH. Review on research and application of stainless steel reinforced concrete. In: MATEC Web of Conferences. Vol. 63. EDP Sciences; 2016. p. 03003

[4] Manning DG. Corrosion performance of epoxy-coated reinforcing steel: North American experience. Construction and Building Materials. 1996;**10**(5):349-365

[5] El-Hacha R, Mirmiran A, Cook A, Rizkalla S. Effectiveness of surface-applied corrosion inhibitors for concrete bridges. Journal of Materials in Civil Engineering. 2011;**23**(3):271-280

[6] Stankiewicz A, Szczygieł I, Szczygieł B. Self-healing coatings in anti-corrosion applications. Journal of Materials Science. 2013;**48**(23):8041-8051

[7] Giannaros P, Kanellopoulos A, Al-Tabbaa A. Sealing of cracks in cement using microencapsulated sodium silicate. Smart Materials and Structures. 2016;**25**(8):084005

[8] ACI Committee 440.1R-15. Guide for the Design and Construction of Structural Concrete Reinforced With Firber-Reinforced Polymer (FRP) Bars (ACI440.1R-15). 2015

[9] Gudonis E, Timinskas E, Gribniak V, Kaklauskas G, Arnautov AK, Tamulėnas V. FRP reinforcement for concrete structures: State-of-the-art review of application and design. Engineering Structures and Technologies. 2013;**5**(4): 147-158

[10] Nanni A, Dolan CW. Fibre-reinforced-plastic (FRP) reinforcement for concrete structures. In: Properties and Application, Developments in Civil Engineering. 1993. 248 p

[11] Triantafillou TC, Antonopoulos CP. Design of concrete flexural members strengthened in shear with FRP. Journal of Composites for Construction. 2000; **4**(4):198-205

[12] ACI 440.1R-06. Guide for the Design and Construction of Concrete Reinforced with FRP Bar. Farmington Hills, MI, USA: American Concrete Institute; 2006

[13] Kalpana VG, Subramanian K. Behavior of concrete beams reinforced with GFRP BARS. Journal of Reinforced Plastics and Composites. 2011;**30**(23): 1915-1922

[14] Goldston M, Remennikov A, Sheikh MN. Experimental investigation of the behaviour of concrete beams reinforced with GFRP bars under static and impact loading. Engineering Structures. 2016; **113**:220-232

[15] Mufti AA, Onofrei MB, Benmokrane B, Banthia N, Boulfiza M, Newhook JP, et al. Field study of glass-fibre-reinforced polymer durability in concrete. Canadian Journal of Civil Engineering. 2007;**34**(3):355-366

[16] Goldston MW, Remennikov A, Sheikh MN. Flexural behaviour of GFRP reinforced high strength and ultra high strength concrete beams. Construction and Building Materials. 2017;**131**:606-617

[17] Krasniqi C, Kabashi N, Krasniqi E, Kaqi V. Comparison of the behavior of

GFRP reinforced concrete beams with conventional steel bars. Pollack Periodica. 2018;**13**(3):141-150

[18] Theriault M, Benmokrane B. Effects of FRP reinforcement ratio and concrete strength on flexural behavior of concrete beams. Journal of Composites for Construction. 1998;**2**(1):7-16

[19] Gholizadeh S, Leman Z, Baharudin BH. A review of the application of acoustic emission technique in engineering. Structural Engineering and Mechanics. 2015;**54**(6):1075-1095

[20] Ohno K, Ohtsu M. Crack classification in concrete based on acoustic emission. Construction and Building Materials. 2010;**24**(12): 2339-2346

[21] Sharma A, Sharma S, Sharma S, Mukherjee A. Investigation of deterioration in corroding reinforced concrete beams using active and passive techniques. Construction and Building Materials. 2018a;**161**:555-569

[22] Sharma A, Sharma S, Sharma S, Mukherjee A. Monitoring invisible corrosion in concrete using a combination of wave propagation techniques. Cement and Concrete Composites. 2018b;**90**:89-99

[23] Garhwal S, Sharma S, Sharma SK. Monitoring the flexural performance of GFRP repaired corroded reinforced concrete beams using passive acoustic emission technique. Structural Concrete. 2021;**22**(1):198-214

[24] Ohtsu M, Mori K, Kawasaki Y. Corrosion process and mechanisms of corrosion-induced cracks in reinforced concrete identified by AE analysis. Strain. 2011;**47**:179-186

[25] Prem PR, Murthy AR. Acoustic emission monitoring of reinforced concrete beams subjected to four-point-bending. Applied Acoustics. 2017;**117**: 28-38

[26] Dunn SE, Young JD, Hartt WH, Brown RP. Acoustic emission characterization of corrosion induced damage in reinforced concrete. Corrosion. 1984;**40**(7):339-343

[27] Aggelis DG, Soulioti D, Barkoula NM, Paipetis AS, Matikas TE, Shiotani T. Acoustic emission monitoring of steel-fiber reinforced concrete beams under bending. Journal of Acoustic Emission. 2008;**14**:287-294

[28] Aggelis DG, De Sutter S, Verbruggen S, Tsangouri E, Tysmans T. Acoustic emission characterization of damage sources of lightweight hybrid concrete beams. Engineering Fracture Mechanics. 2019;**210**:181-188

[29] Sharma G, Sharma S, Sharma SK. Non-destructive evaluation of steel and GFRP reinforced beams using AE and DIC techniques. Structural Engineering and Mechanics. 2021;**77**(5):637-650

[30] Yuan Q, Shi C, De Schutter G, Audenaert K, Deng D. Chloride binding of cement-based materials subjected to external chloride environment–a review. Construction and Building Materials. 2009;**23**(1):1-3

[31] IS 10262. Indian Standard Code for Concrete Mix Proportioning Guidelines (First Revision). New Delhi, India: Bureau of Indian Standards; 2009

[32] IS 10262. Indian Standard Code for Concrete Mix Proportioning Guidelines (First Revision). New Delhi India: Bureau of Indian Standards; 2008

[33] ASTM D7205. Standard test Method for Tensile Properties of Fiber Reinforced Polymer Matrix Composite Bars. West Conshohocken, Pennsylvania, USA: ASTM International; 2006

[34] ACI 319. Building Code Requirements for Structural Concrete and Commentary. Farmington Hills, MI, USA: American Concrete Institute; 2019

Chapter 3

Application of Acoustic Emissions Technique in Assessment of Cracking Performance of Asphalt Pavement Materials

Behzad Behnia

Abstract

This chapter focuses on various applications of acoustic emissions (AE) technique in evaluation of cracking in asphalt pavements including (1) assessment of low-temperature cracking of asphalt binders and mixtures and (2) quantitative characterization of rejuvenators' efficiency in restoring aged asphalt pavements to their crack-resistant state. The AE-based embrittlement temperature results of 24 different asphalt materials consisting of eight different binders, each at three oxidative aging levels are presented. Results show that embrittlement temperatures correlated well with corresponding bending beam rheometer (BBR-based) critical cracking temperatures with R^2 = 0.85. This chapter also presents application of AE for evaluation of rejuvenators' efficiency on asphalt materials at various oxidative aging levels. The Geiger's iterative source location method was employed to accurately determine embrittlement temperatures throughout the thickness of rejuvenator-treated asphalt samples. Results showed that the low temperature cracking properties of oxidative aged materials after 2 weeks of dwell time of rejuvenator have been recuperated. Moreover, it was observed that cracking characteristics of aged asphalt 6–8 weeks after applying rejuvenator far exceeded that of the virgin materials. The promising results suggest that the AE technique can be considered as a viable approach for the assessment of low temperature behavior of asphalt pavements.

Keywords: acoustic emission, thermal cracking, asphalt pavements, bending beam rheometer, source location, rejuvenator, oxidative aging

1. Introduction

As an important component of the transportation infrastructure, asphalt pavement is composed of multi-layer complex system of different materials subjected to various combinations of traffic as well as environmental loadings. During their lifetime asphalt pavements experience various forms of distresses as they undergo oxidative aging, freeze-thaw cycles, and traffic repetitions. One of the most widespread type of deterioration in asphalt roads which shortens the pavement life and results in premature failure of the pavement structure is cracking. An accurate and realistic assessment of cracking performance of asphalt pavements has remained as a challenging task for civil engineers.

The AE technique has become very popular in recent decades due to its unique ability in detecting and locating microstructural failures in different types of materials. This method has been successfully used for damage detection of various materials such as steel, concrete, wood, and rock. However, for the case of asphalt roads, there has been only a limited application of this technology in damage assessment of asphalt pavements. In one of the studies Khosla and Goetz [1] used the acoustic emission approach at low temperatures to locate crack initiation and propagation in indirect tensile (IDT) asphalt concrete specimens. They found that the material failure due to fracture was accompanied by a sudden increase in total AE counts where a big portion of AE counts occurred at about 80% of the peak load. In another study conducted by Valkering and Jongeneel [2], AE technique was implemented to carefully monitor the thermally-induced cracking in asphalt concrete materials subjected to low temperature cooling cycles (−10°C to −40°C). Results showed that the AE activity of the material such as number of events were strongly correlated with the extent of thermal fracture in the specimens. Results also demonstrated that at low temperatures the source of AE activities in restrained specimens were crack initiation originated from defects exists in the material. In the research study performed by Hesp et al. [3] the AE method was employed for restrained asphalt concrete specimens at low temperatures (−32°C to −20°C) to measure and to detect crack initiation and propagation in the restrained samples. They compared the total amount of AE activities in different mixtures and found that the styrene-butadiene-styrene (SBS)-modified asphalt concrete materials exhibited less AE activities as compared to that of for the unmodified asphalt concrete mixes. The AE approach was implemented by Li et al. [4–8] to evaluate and to quantify fracture in semi-circular bending (SCB) asphalt specimens at −20°C. They concluded that most of the AE activities in the material happens at about 70% of the material strength. Their results also showed that the maximum intensity of AE amplitudes correlated well with the extent of macrocracking damage in the specimen. They also found that the location of AE events is the good indicator of approximate size of the fracture process zone (FPZ). Nesvijski and Marasteanu [9, 10] in another research study, used the AE spectral analysis approach at low temperatures in order to investigate and assess fracture in semi-circular bending (SCB) asphalt samples. They were able to successfully demonstrate that the AE approach could be applied for accurate characterization of cracking in asphalt concrete materials.

This chapter will focus on various applications of the acoustic emission technique in asphalt pavements including: (1) assessing the low-temperature cracking performance of asphalt binders and asphalt pavement materials (2) use of acoustic emission technique for quantitative evaluation of restoring aged asphalt pavements with rejuvenators.

2. Implementation of AE approach for low-temperature cracking assessment of asphalt binders and asphalt concrete materials

Low temperature cracking, a.k.a. thermal cracking, is a very common type of damage occurring in asphalt pavements located either in regions with cold climates or in milder climate regions with large daily temperature fluctuations. In asphalt pavements built in cold climates with severe winters, thermal cracking usually happens as a result of fast cooling rates (single-event thermal cracking). On the other hand, in asphalt roads located in regions with milder climate, thermal cracks develop at a slower rate, and it usually takes several cooling cycles for cracks to initiate and propagate through the pavement thickness (thermal fatigue cracking) [11]. When the temperature drops, surface of the pavement has the lowest temperature, and the temperature changes are highest there. Thermal tensile stresses develop in the restrained pavement layer due to the change in pavement temperature. The thermally-induced stresses are greatest in the longitudinal direction of the road which will lead to formation of transversely-oriented surface-initiated thermal cracks of various lengths and widths along the road.

Numerous research studies have demonstrated that the low temperature characteristics of asphalt pavements are closely related to that of the asphalt binder used in pavement construction. The AE method is implemented to evaluate the thermal cracking in asphalt binders. The AE binder sample consists of a 6 mm thick layer of asphalt binder bonded to an aluminum plate. To conduct the test, prepared specimens are placed inside the freezer and exposed to decreasing temperatures, ranging from 20°C to −40°C, or even to −50°C, if necessary for some polymer modified binders. To continuously monitor and record the sample temperature, a K-type thermocouple is placed on the specimens' surface. Due to the relatively small size of the AE sample, there is a thermal lag at the beginning of the test, which becomes negligible at temperature lower than −10°C. Differential thermal contraction between aluminum and asphalt binder induces progressively higher thermal stresses in the binder leading to formation of thermal cracks in the material. Thermal cracks formation in the sample is accompanied by a release of elastic energy in the form of transient waves which could be picked up using the AE piezoelectric sensor(s) mounted on aluminum plate. The critical cracking temperature, a.k.a. the embrittlement temperature, of the asphalt binders tested are determined by processing and analyzing the emitted elastic waves captured during the tests using the AE technique. **Figure 1a** schematically illustrates an AE testing sample of asphalt binder with an aluminum substrate [12–21].

To conduct the AE test for asphalt concrete materials, a semicircular-shaped asphalt concrete sample with a 50°mm thickness and a 150°mm diameter is used as the testing specimen, see **Figure 1b**. The testing sample for asphalt concrete can be fabricated from either field cores or from gyratory compacted samples. To conduct the AE test, similar to the binder test, the prepared AE sample is subjected to decreasing temperatures ranging from 20°C to -40°C and the acoustic activities and temperature of asphalt material test sample is continuously monitored and recorded using piezoelectric AE sensors and a K-type thermocouple, respectively. The source of acoustic emission activities in asphalt concrete materials is formation of thermally-induced microdamages within the asphalt mastic. As a heterogeneous viscoelastic material, the thermally induced stresses develop in asphalt concrete due to the thermal contraction mismatch between aggregates and surrounding asphalt mastic [12–26].

The acoustic emission testing set up used for assessing asphalt materials consists of several wideband piezoelectric AE sensors along with pre-amplifiers and data acquisition system with processing and analysis software. The Digital Wave-Model B1025 wideband AE sensors with nominal frequency range of 20 kHz to 1.5 MHz used in this study in order to record and continuously monitor AE activities of the material while conducting the experiment. To reduce extraneous noise, the AE signals picked up by AE sensors are first pre-amplified 20 dB using broad-band pre-amplifiers. Then AE signals are amplified again 21 dB for a total of 41 dB. At the end signals are filtered using a 20 kHz high-pass double-pole filter through using the signal conditioning unit. A 16-bit analog to digital converter (ICS 645B-8) with 2 MHz sampling frequency and a length of 2048 points per channel per acquisition trigger are used to digitize the signals and outputs are stored for the post-processing.

In general there are two methods normally used to analyze AE signals: (1) the "classic" or "parameter-based" method; (2) the "quantitative" or "signal-based" approach. In the first approach, the AE signals are not recorded, instead only some

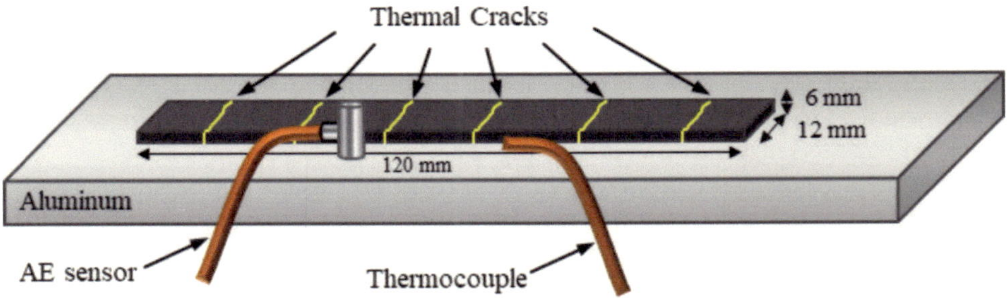

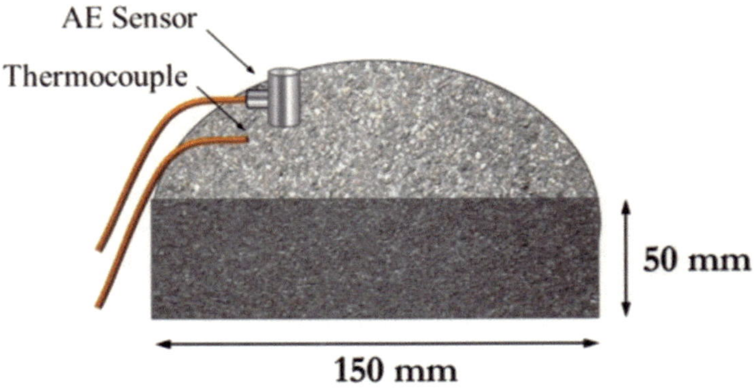

Figure 1.
AE testing samples used for (a) asphalt binders (b) asphalt concrete materials.

AE parameters are recorded and analyzed. Whereas in the signal-based approach, the actual AE signals are recorded and used to analyze the materials microstructure. The failure and microdamage occurring in the material could generate significant number of AE signals within a very short time generating big amount of AE data. In the parameter-based method only some rudimentary analysis can be performed on AE data however it is faster than the quantitative method. On the other hand, while the signal-based approach is slower, it is capable of more sophisticated analysis of performance of the material.

For evaluation of thermal cracking in asphalt materials, both parameter-based and signal-based techniques were implemented on recorded AE signals and associated test temperature. AE event is an individual waveform with the threshold of 0.1 V and the energy level equal to or greater than 4 V^2 μs. The emitted energy associated with each event is one of the important characteristics of an AE signal and can be calculated using Eq. (1), where E_{AE} is the AE energy of an event (V^2 μsec) with duration of time t (μsec) and recorded voltage of V(t) [1].

$$E_{AE} = \int_0^t V^2(t)dt \tag{1}$$

Figure 2 shows a typical plot of AE events counts versus temperature for typical asphalt binder and asphalt concrete AE tests which consists of four distinct regions, namely: (1) pre-cracking, (2) transition, (3) stable cracking, and (4) fully cracked regions. In the "pre-cracking region", thermally-induced stresses in the sample are building up and they are still below the strength of the material. As a result no damage and consequently no AE events are observed within this region. In the "transition" region, as soon as the thermal stresses reach the strength of the material, microdamages form in the material which manifests itself as a cluster of high amplitude AE events. The temperature corresponding to the AE event with the first peak energy within the transition region has been termed the "embrittlement temperature," as shown in **Figure 3**. The embrittlement temperature is the onset of damage in asphalt material. Results has demonstrated that the embrittlement temperature is a fundamental material state which is independent of material constraint, sample size (as long as a statistically representative volume or larger is used), and sample shape [15]. In the "transition region", material behavior gradually changes from a quasi-brittle to a brittle state where resistance to fracture is generally very low, allowing microdamages to propagate readily.

The third region is the "stable cracking region" which normally initiates at a very low temperatures when the material is brittle. Significant amount of AE activities are observed during this region. The last region, is the "fully cracked region" where the rate of AE activities of the sample begins to reduce until it reaches almost zero at the end of this region. The AE activities originate from formation of new micro-damage inside the sample. Thus reduction in the rate of AE activity can be linked to the presence of plenty of microdamage in the sample. This region is usually observed when the sample is cooled down to very cold temperatures allowing all microdamage to develop within the sample [15].

Figure 4 illustrates the typical envelope locus of AE event energies of asphalt samples and demonstrates the intensity of the released energies of AE events. In the pre-cracking region, the envelope locus is zero and suddenly at the beginning of the transition region it jumps to its maximum magnitude. The magnitude of AE event energies gradually tapers off in stable cracking region until it reaches almost zero in the fully cracked region.

The histogram presented in **Figure 5** shows the graphical representation of the distribution of AE events energies for asphalt materials. Results suggest that only a small portion of AE events are high energy events while the rest of the events are in fact low energy. Generally, the energy content of an event is proportional with the size of the microdamage causing that event. The high energy events result from the formation of large microcracks while the low energy AE events could be linked to formation of hairline microcracks in the material.

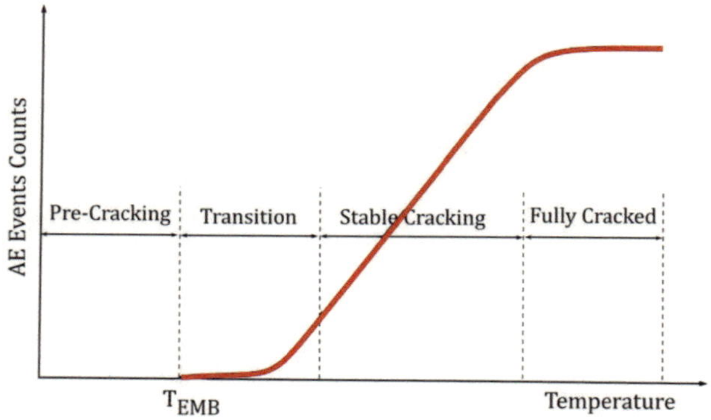

Figure 2.
Typical AE event counts vs. temperature plot regions.

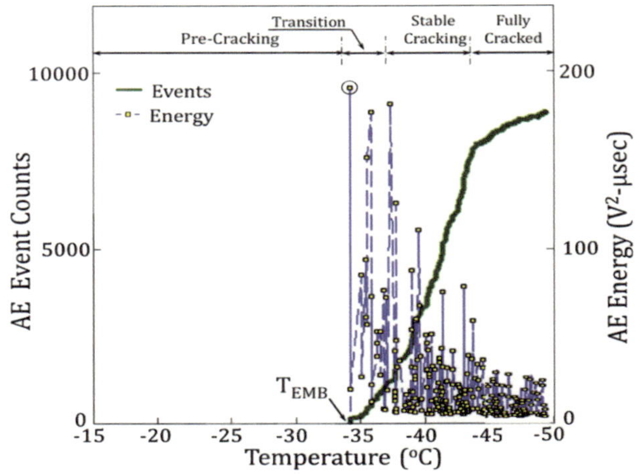

Figure 3.
Typical plot of event count and AE energy vs. temperature [15].

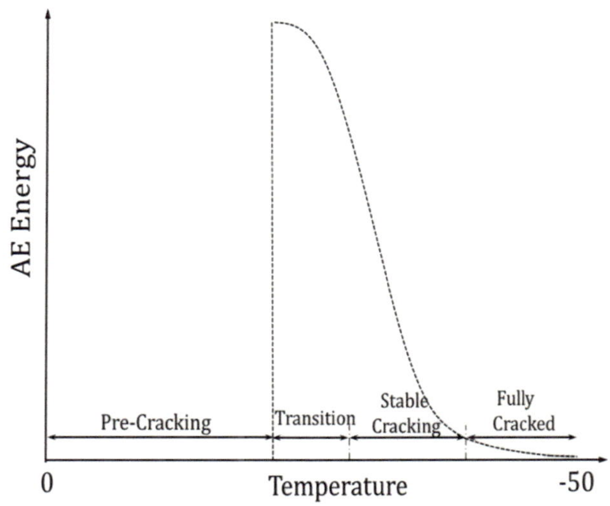

Figure 4.
Typical envelope locus of AE events energy during thermal cooling [15].

The AE test results for 24 different types of asphalt materials (eight different binders, each at three aging levels) including: AAA-1 (PG 58-28), AAB-1 (PG 58-22), AAC-1 (PG 58-16), AAD-1 (PG 58-28), AAF-1 (PG 64-10), AAG-1 (PG 58-10), AAK-1 (PG 64-22), AAM-1 (PG 64-16) are presented in **Figure 6**. In this experiment each binder was tested at three aging levels: (1) unaged (TANK) (2) short-term aged (RTFO), and (3) long-term aged (PAV). The ASTM D2872-04 (ASTM 2004) and ASTM D6521-05 (ASTM 2008) were used to perform the oxidative aging process of RTFO and PAV binders, respectively. It should be mentioned that in PG XX-YY used for expressing the Performance Grade of asphalt materials, XX corresponds to the expected average high temperature of asphalt pavement over a 7 days, and YY is the lowest expected temperature of the pavement.

Results show that the AE embrittlement temperatures correlated well with the bending beam rheometer (BBR-based) critical cracking temperatures with $R^2 = 0.85$. Results suggest that AE-based embrittlement temperatures are lower than the corresponding BBR-based critical cracking temperatures. This could be

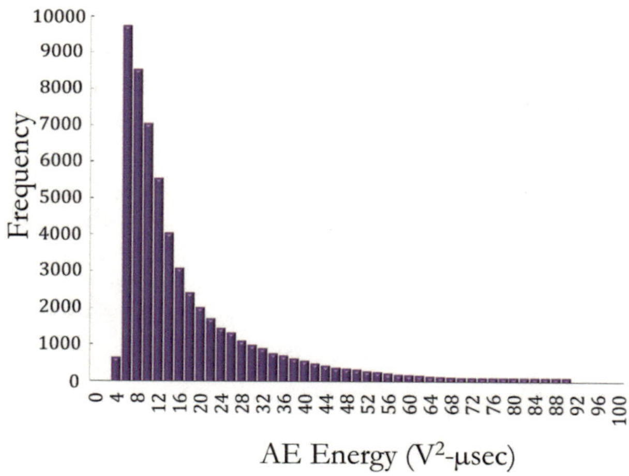

Figure 5.
Typical histogram of AE events energies for asphalt binder material [15].

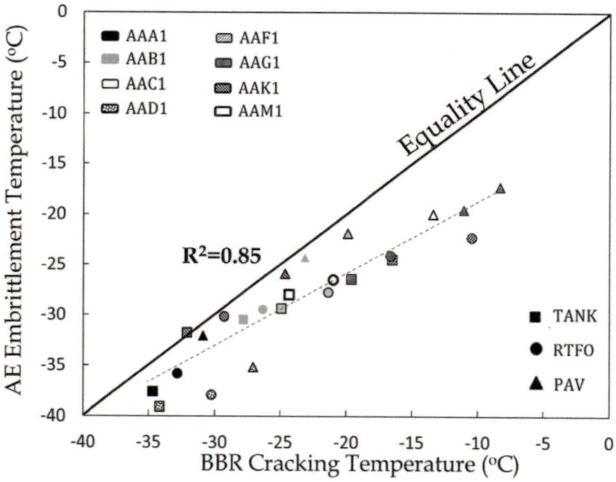

Figure 6.
Correlation between AE embrittlement temperature and BBR-based cracking temperature illustrating the conservative nature of the BBR-based cracking temperatures [21].

attributed to the fact that the AE-based embrittlement temperatures are directly related to the cracking performance of the material while the BBR- base critical temperatures are based upon the binder's rheological material properties and include an inherent factor of safety to avoid low-temperature pavement cracking. In addition, numerous studies have demonstrated that AE approach is sensitive to aging level of the material and could successfully evaluate asphalt materials at different oxidative aging levels. Finding of different studies show that the embrittlement temperature of asphalt materials is sensitive to aging levels, where $T_{EMB\text{-}TANK} < T_{EMB\text{-}FTFO} < T_{EMB\text{-}PAV}$.

3. Use of acoustic emission technique for quantitative evaluation of restoring aged asphalt pavements with rejuvenators

Oxidative aging is a common problem in asphalt pavements which leads to an increase in stiffness and loss of ductility and cohesion of binders. It negatively affects the fracture resistance of pavements. Certain chemical properties of the asphalt binders such as asphaltenes to maltenes ratio changes in the oxidation process. The oxidation rate of asphalt materials is accelerated at high temperatures and/or high exposure to ultraviolet light and air [21–26]. Different methods such as pavement surface milling and the application of rejuvenators are employed to restore asphalt pavements to their crack-resistant state. Application of rejuvenators is one of the popular techniques to restore the physical and chemical properties of aged asphalt materials. Rejuvenators change the asphaltenes to maltenes ratio to its original state leading to softening the aged asphalt materials [21–26]. Rejuvenators are generally sprayed on the surface of aged pavements. It is very important that rejuvenator could penetrate the surface via capillary action and gravity and diffuse through the aged asphalt.

Currently there is no standardized method to assess the performance of rejuvenators when applied in the field. The efficiency of rejuvenators is evaluated by the following three methods which are cumbersome and time consuming and they are not often used.: (1) estimating the penetration of rejuvenator in the pavement by comparing the penetration value of the binder at 25°C in the asphalt binder extracted from untreated and treated sample; (2) comparing the asphalt binders' viscosity at 60°C obtained from untreated and treated cores; and (3) comparing the amount of loss in aggregates in the abrasion test in untreated vs. treated samples [26].

The AE source location approach has recently been employed to assess the efficiency of rejuvenators in restoring aged asphalt materials to their original crack resistant condition. The Geiger's iterative source location method was used to accurately detect the source of AE activities in the material [26, 27]. This iterative technique is based on the Gauss-Newton algorithm. To build the arrival time function of the ith sensor, see Eq. (2), data from at least four sensors is required for the Geiger's method:

$$f_i(x, y, z, t) = T_s + \frac{1}{v}\sqrt{(x_i - X_s)^2 + (y_i - Y_s)^2 + (z_i - Z_s)^2} \tag{2}$$

where (X_s, Y_s, Z_s) represent the spatial coordinates of the AE source, (x_i, y_i, z_i) is the coordinates of the ith sensor, v is the velocity of wave in the material, t_i and T_s represent the known receiving time and unknown AE source event occurring time by the ith sensor, respectively. Taylor series is used to expand Eq. (2) at a point (x_0, y_0, z_0), close to the actual source leading to Eq. (3):

$$f_i(x, y, z, t) = f_i(x_0, y_0, z_0, t_0) + \epsilon_i \tag{3}$$

where ϵ_i is the residual term, a.k.a. the correction vector, which is the difference between the calculated arrival time and the observed arrival time with respect to the ith sensor. The correction vector can be determined using the first order derivatives of the arrival time function. The Geiger's method tries to minimize the correction vector by going through several iterations of Eq. (4).

$$\epsilon_i = \frac{\partial f_i}{\partial x}\delta x + \frac{\partial f_i}{\partial y}\delta y + \frac{\partial f_i}{\partial z}\delta z + \frac{\partial f_i}{\partial t}\delta t \tag{4}$$

In this chapter results from one of the studies on evaluation of rejuvenators on aged asphalt materials are presented where PG64-22 was used as the based binder. The asphalt content of the mixture was 5.6% by weight and the gyratory compacted specimens were made using a maximum aggregate size of 19 mm. Some specimens were aged in the oven for 2 h at 155°C to simulate the aging level during plant production. Part of the specimens were aged in the oven for 36 h at 155°C (in addition to the short term aging) to mimic the long term aged asphalt pavement materials. The oxidative aging process was done on loose mixtures in order to obtain uniformly-aged compacted samples. **Figure 7** shows one of specimens with eight AE sensors mounted on the top and bottom surfaces of the specimen, four sensors on each side. To avoid numerical instability, AE sensors pattern at the bottom of the specimen has a 45° offset angle with respect to the pattern of sensors coupled on the top surface.

Some aged specimens were treated by spraying a thin layer of rejuvenator on the top surface of the sample. The amount of rejuvenator used was 10% by weight

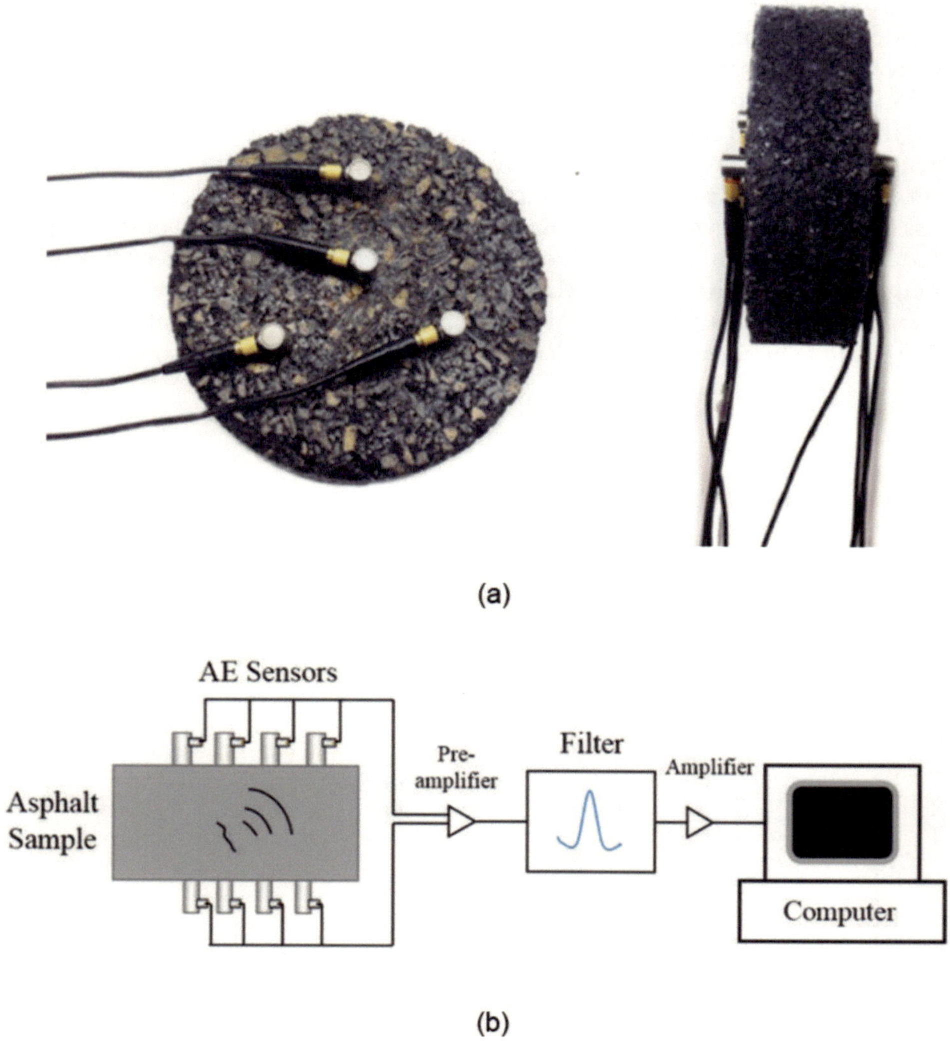

Figure 7.
(a) Oxidative aged AE testing sample with eight piezoelectric sensors, four sensors on each side. (b) AE testing setup used for source location [21].

of the asphalt binder. The rejuvenator-treated specimens were then stored for a prescribed dwell time of 2, 4, 6, and 8 weeks before performing the AE tests. After each dwell time, specimens were tested using the same AE source location procedure used to test the 36 h and 2 h aged specimens, allowing the estimation of the embrittlement temperatures throughout the sample thickness.

To characterize the efficiency of rejuvenator on the aged asphalt materials, the embrittlement temperatures of the material were determined throughout the thickness of asphalt concrete samples by implementing the AE source location method. **Figure 8** illustrates the embrittlement temperatures results vs. sample thickness for different aged asphalt concrete materials. The effect of oxidative aging on the embrittlement temperature is clearly noticeable as the embrittlement temperature of the short-term aged sample (−22°C) is lower than that of the 36 h aged samples (−13°C). It is also observed that for all specimens the embrittlement temperatures of oxidative aged materials after 2 weeks of dwell time of rejuvenator have been recuperated. The test results obtained from samples after 6 and 8 weeks of dwell time were quite surprising as the embrittlement temperatures of the aforementioned samples far exceeded the embrittlement temperatures of the virgin materials. Moreover, the method was also able to successfully capture the embrittlement temperature gradation throughout the sample thickness for the dwell times of 2 and 4 weeks. This could be attributed to the fact that the rejuvenator has had enough time to penetrate and act on the top material layers. Results suggest that the AE method can be employed to accurately evaluate the graded embrittlement temperature properties of oxidative aged asphalt pavements. One important outcome of this study is that the AE approach can be used to intelligently select the best maintenance strategies for oxidative aged asphalt roads through optimizing the amounts of rejuvenators required to restore pavement to the original crack-resistant condition, or by optimizing the relative amount of milling and surface replacement of asphalt roads. In addition, the AE results obtained from source location approach were found to be consistent with those of obtained from non-collinear ultrasonic wave mixing method [28–31].

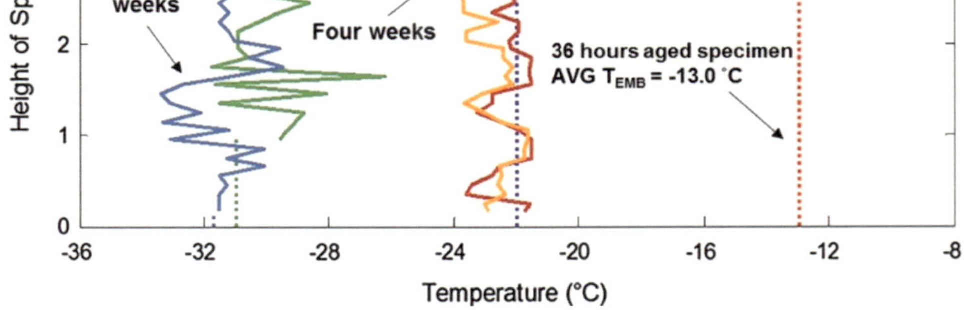

Figure 8.
Average measured embrittlement temperatures of rejuvenator-treated oven-aged asphalt concrete samples (for 36 h at 135°C) after dwell times of 2, 4, 6 and 8 weeks [21, 25, 26].

4. Concluding remarks

The acoustic emission approach has been successfully implemented to address the shortage for accurate and reliable techniques to evaluate cracking performance of asphalt pavements. In addition, the AE method has been applied for assessing the efficiency of rejuvenator-treated pavements and to evaluate and characterize virgin, short-term, and long-term asphalt binders and asphalt concrete materials. The AE technique has also been employed in different areas such as evaluating asphalt pavements containing recycled materials such as RAP or RAS, assessing the effect of cooling cycles upon the structural integrity of pavements, and characterizing the thermal cracking performance of graded, i.e., aged asphalt pavements. The promising results from aforementioned studies suggest that the AE technique can be considered as a viable approach for the assessment of asphalt pavements. Moreover, when used for preventive maintenance and rehabilitation, AE method can serve as a powerful tool in enhancing pavement sustainability. Both up-stream and down-stream suppliers and producers of asphalt concrete binders could benefit from AE technique. The up-stream supplies of polymer, chemical, and other additives (warm-mix additives, antistrip agents) could use AE for rapid assessment of low-temperature characteristics of trial formulations, and could quickly evaluate the compatibility of blended additive systems. Asphalt mixture designers could take advantage of the AE technology in order to verify binder grade selection and to optimize the amount of recycled materials used in the pavement. Finally, the pavement owners could implement AE for quality assurance of asphalt binders and asphalt mixtures, for periodic pavement condition assessments, and for the scheduling of preventive maintenance and rehabilitation, where pavement cracking is of concern.

Author details

Behzad Behnia
Department of Civil and Environmental Engineering, Clarkson University, Potsdam, NY, USA

*Address all correspondence to: bbehnia@clarkson.edu

References

[1] Khosla NP, Goetz WH. Tensile characteristics of bituminous mixtures as affected by modified binders. Joint Highway Research Project. Indiana, USA: Purdue University and the Indiana State Highway Commission; 1979

[2] Valkering CP, Jongeneel DJ. Acoustic emission for evaluating the relative performance of asphalt mixes under thermal loading conditions (with discussion). Journal of the Association of Asphalt Paving Technologists. 1991;**60**:160-187

[3] Hesp S, Terlouw T, Vonk W. Low temperature performance of SBS-modified asphalt mixes. Asphalt Paving Technology. 2000;**69**:540-573

[4] Li X, Marasteanu M. Evaluation of the low temperature fracture resistance of asphalt mixtures using the semi circular bend. In Association of Asphalt Paving Technologists-Proceedings of the Technology, AAPT. 2004;**73**:401-426

[5] Li X, Marasteanu MO. Investigation of low temperature cracking in asphalt mixtures by acoustic emission. Road materials and Pavement Design. 2006;7(4):491-512

[6] Li X, Marasteanu MO, Iverson N, Labuz JF. Observation of crack propagation in asphalt mixtures with acoustic emission. Transportation Research Record: Journal of the Transportation Research Board. 2006;**1970**(1):171-177

[7] Li X, Marasteanu M, Turos M. Study of low temperature cracking in asphalt mixtures using mechanical testing and acoustic emission methods. In Asphalt Paving Technology: Association of Asphalt Paving Technologists-Proceedings of the Technical Sessions. Association of Asphalt Paving Technologist. December 2007;**76**:427-453

[8] Li X, Marasteanu M. The fracture process zone in asphalt mixture at low temperature. Engineering Fracture Mechanics. 2010;**77**(7):1175-1190

[9] Nesvijski E, Marasteanu M. Wavelet transform and its applications to acoustic emission analysis of asphalt cold cracking. Journal of Nondestructive Testing & Ultrasonics. 2007;**12**:1-13.

[10] Nesvijski E, Marasteanu M. Spectral analysis of acoustic emission of cold cracking asphalt. Journal of Nondestructive Testing & Ultrasonics. 2006;**11**(10)

[11] Kim YR. Modeling of Asphalt Concrete. McGraw-Hill Construction. New York, McGraw-Hill: ASCE Press; 2008. ISBN-13: 978-0071464628

[12] Buttlar WG, Behnia B, Reis HM. An acoustic emission-based test to determine asphalt binder and mixture embrittlement temperature. NCHRP IDEA Project No. 144. Washington, D.C. USA: Transportation Research Board; 2011

[13] Behnia B, Buttlar WG, Reis H, Apeagye AK. Determining the embrittlement temperature of asphalt binders using an acoustic emission approach. In: Proceedings of Structural Materials Technology Conference in New York; August 2010

[14] Behnia B, Dave E, Ahmed S, Buttlar W, Reis H. Effects of recycled asphalt pavement amounts on low-temperature cracking performance of asphalt mixtures using acoustic emissions. Transportation Research Record: Journal of the Transportation Research Board. 2011;**2208**:64-71

[15] Behnia B. An acoustic emission-based test to evaluate low temperature behavior of asphalt materials [Doctoral dissertation]. Champaign: University of Illinois at Urbana; 2013

[16] Behnia B, Dave EV, Buttlar WG, Reis H. Characterization of embrittlement temperature of asphalt materials through implementation of acoustic emission technique. Construction and Building Materials. 2016;**111**:147-152

[17] Behnia B, Buttlar WG, Reis H. Nondestructive low-temperature cracking characterization of asphalt materials. Journal of Materials in Civil Engineering. 2016;**29**(5):04016294

[18] Behnia B, Buttlar WG, Reis H. Nondestructive evaluation of thermal damage in asphalt concrete materials. Journal of Testing and Evaluation, ASTM. 2017;**45**(6):1948-1958

[19] Behnia B, Reis H. Self-healing of thermal cracks in asphalt pavements. Construction and Building Materials. 2019;**218**:316-322

[20] Hill B, Oldham D, Behnia B, Fini EH, Buttlar WG, Reis H. Evaluation of low temperature viscoelastic properties and fracture behavior of bio-asphalt mixtures. International Journal of Pavement Engineering. 2018;**19**(4):362-369

[21] Behnia B, Buttlar WG, Reis H. Evaluation of low-temperature cracking performance of asphalt pavements using acoustic emission: A review. Applied Sciences. 2018;**8**(2):306

[22] Hakimzadeh S, Behnia B, Buttlar WG, Reis H. Implementation of nondestructive testing and mechanical performance approaches to assess low temperature fracture properties of asphalt binders. International Journal of Pavement Research and Technology. 2017;**10**(3):219-227

[23] ME MG, Behnia B, Buttlar WG, Reis H. Use of nonlinear acoustic measurements for estimation of fracture performance of aged asphalt mixtures. Transportation Research Record. 2017;**2631**(1):11-19

[24] Sun Z, Farace N, Arnold JW, Behnia B, Buttlar WG, Reis H. Quantitative evaluation of rejuvenators to restore embrittlement temperatures in oxidized asphalt mixtures using acoustic emission. In: Health Monitoring of Structural and Biological Systems. International Society for Optics and Photonics; Proceedings of the SPIE; 2015. Volume 9438, id. 94381C, 14 pp. DOI: 10.1117/12.2082573

[25] Sun Z, Farace N, Behnia B, Buttlar WG, Reis H. Quantitative evaluation of rejuvenators to restore embrittlement temperatures in oxidized asphalt mixtures using acoustic emission. Journal of Acoustic Emission. 2015;**33**:1-15

[26] Sun Z, Behnia B, Buttlar WG, Reis H. Acoustic emission quantitative evaluation of rejuvenators to restore embrittlement temperatures to oxidized asphalt mixtures. Construction and Building Materials. 2016;**126**:913-923

[27] Geiger L. Probability method for the determination of earthquake epicenters from the arrival time only. Bulletin of St. Louis University. 1912;**8**:60-71

[28] Hill B, Behnia B, Buttlar WG, Reis H. Evaluation of warm mix asphalt mixtures containing reclaimed asphalt pavement through mechanical performance tests and an acoustic emission approach. Journal of Materials in Civil Engineering. 2013;**25**(12): 1887-1897

[29] Hill B, Behnia B, Hakimzadeh S, Buttlar WG, Reis H. Evaluation of low-temperature cracking performance of warm-mix asphalt mixtures. Transportation Research Record. 2012; **2294**(1):81-88

[30] Hill B, Oldham D, Behnia B, Fini EH, Buttlar WG, Reis H. Low-temperature performance characterization of biomodified asphalt mixtures that

contain reclaimed asphalt pavement. Transportation Research Record. 2013;**2371**(1):49-57

[31] Arnold JW, Behnia B, ME MG, Hill B, Buttlar WG, Reis H. Quantitative evaluation of low-temperature performance of sustainable asphalt pavements containing recycled asphalt shingles (RAS). Construction and Building Materials. 2014;**58**:1-8

Chapter 4

Defect Detection in Delaminated Glass-Fibre/Epoxy Composite Plates Using Local Defect Resonance Based Vibro-Thermography Technique

Subhankar Roy and Tanmoy Bose

Abstract

In the present scenario, composites are widely used for various applications in the field of aerospace, automobile, marine, sports, construction and electrical industries. The need of damage inspection for these composite structures has been of great importance. Complicated defects like delaminations present in the composite laminates can be detected effectively using nonlinear acoustic wave spectroscopy (NAWS). One of the NAWS techniques of detecting the delamination is based on intensification of vibration amplitudes at the delamination location, known as local defect resonance (LDR) technique. In this chapter, a numerical investigation for detecting delamination in glass fibre reinforced polymer (GFRP) composite based on vibrothermography technique will be discussed. A single periodic LDR frequency excitation is used to excite the GFRP plate, resulting in a local temperature rise at delamination region due to frictional heating at the damage interface. An explicit dynamic temperature displacement analysis will be carried out for a specific time period of LDR excitation. Subsequently, a heat transfer analysis will be performed to observe the temperature difference at top surface of the delaminated GFRP plate. Thus a numerical investigation will be carried out based on LDR excitation for high contrast imaging of delamination in composite materials using vibro-thermography.

Keywords: local defect resonance (LDR), delamination, vibro-thermography, frictional heating, glass fibre reinforced polymer (GFRP)

1. Introduction

In recent times, composite materials finds a wide variety of applications in several fields of engineering such as automobile, aerospace, building materials, etc. because of their good strength to weight ratio, higher resistance to impact, better fatigue property and lower costs. The composite materials are generally classified into three vast categories according to their matrix material type, namely, polymer matrix composites (PMCs), metal matrix composites (MMC) and ceramic matrix composites (CMCs). The Glass fibre reinforced polymer (GFRP) and carbon fibre

IntechOpen

reinforced polymer (CFRP) represents the largest class of polymer matrix composites (PMCs) which find numerous applications in the fabrication of aerospace structures, turbine blades, automobile skin, etc. These structures are continuously subjected to various cyclic loads and impacts during the tenure of their service and hence may lead to development of damages in their sub-surface layers, in form of delamination. So, this chapter will discuss a technique of determining the location of such delaminations by performing vibro-thermography based on local defect resonance of the defect. The vibro-thermography technique is an efficient technique for determining the location of damage in a complex structure by exciting the structure at its defect frequency, known as the local defect resonance frequency. Excitation at the defect frequency leads to high amplitude vibration at the defect site resulting in clapping action of the layers, thus, generating a local heat at the defect area. This temperature gradient in the defect area can be easily detected using an IR-camera.

The vibro-thermography technique was implemented by many researchers in the past for carrying out non-destructive evaluation of defects in structures. Vibro-thermography was first introduced by for the detection of subsurface damages in composite structures due to fatigue [1]. They studied the elastic and viscoelastic hysteretic effects by mechanically exciting the specimen and obtaining the thermal patterns. It was observed that the material deformation during the excitation was directly related to the heat generated. The study was further implemented by using resonant vibration to obtain high cyclic stresses [2]. In the year 1996, Rantala et al. excited a sample with mechanical shaker and monitored the sample with an infrared camera to perform a lock-in thermography technique [3]. The high amplitude of vibration led to high temperature signatures at low stress levels which is very good for non-destructive evaluation. Another study used lock-in thermography based on optical heating of defect area for large area inspections [4]. Subsequently, a short pulse sound was used in addition to IR imaging to measure surface temperature as a function of time [5]. This led to efficient detection of subsurface cracks due to enhancement of sonic infrared imaging. Vibro-thermography was found to be an outstanding tool for fast detection of small defects like cracks and delaminations [6]. It was observed that the defect can be detected from any side inspection and not required for all-side inspection. Also, a single excitation location is sufficient to perform a large area scanning and is found very useful for composites with multiple materials having different thermal properties. The vibro-thermography technique was also implemented for determining the size of a defect [7]. In this study, a numerical model was developed using finite difference method where it was observed that the size of defect influences phase angle data, thus affecting the defect depth estimation. In the following year, the lock-in thermography technique was used to differentiate between location, shape and size of defects in case of cracks and corrosion defect [8]. Subsequently, the lock-in thermography was implemented for detection of vertical cracks of arbitrary shape to determine geometry and location of defect [9]. An algorithm is developed to obtain crack shape reconstructions by optimizing the data before entering the algorithm. Deep cracks are precisely detected although the shape of the crack is obtained as rounded and having a slightly over-estimated area. Another group of researchers investigated novel hybrid thermographic techniques in addition to traditional optically excited thermography using external optical radiation such as heaters, flashes and laser systems [10]. Different techniques such as ultrasonic stimulated thermography with ultrasonic waves and damage resonance to enhance the sensitivity of micro-cracks; microwave thermography that uses electromagnetic radiation at microwave frequency bands; and eddy current stimulated thermography to generate induction heating is used for detection of delaminations and cracks. Subsequently, an integrated nonlinear ultrasound and Vibro-thermography based non-destructive testing was proposed using shared excitation sources to solve low sensitivity and small detection area problems in carbon fibre reinforced polymer structures [11]. The study involved time-domain analysis and fast Fourier transform of image sequences obtained from specimens with different impact loads. In case of visible damages, FFT did not improve the sound to noise ratio although the location and shape of the defect were captured in detail. On the other hand, the barely visible impact damages were only detected using nonlinear ultrasound and vibro-thermography technique due to a better sound to noise ratio. Non-destructive testing of CFRP with impact damages using ultrasonic simulation was further implemented using two approaches: low power resonant ultrasonic stimulation and high power stimulation at a fixed frequency [12]. The high power ultrasonic IR thermography using magnetostriction and piezoelectric transmitters proved to be highly informative but at the same time, ultrasonic IR testing needs a higher level of energy consumption in order to induce temperature signatures of desired amplitude. In contrast, low power ultrasonic stimulation using defect resonance allows investigation of complex structured flaws like impact damages while reducing the energy consumption to half when compared with high power ultrasonic thermography. The study was extended by the same group where it is shown that small deviations in the frequency of acoustic signal supplied to transducer from frequency of main resonance leads to a reduction in radiation intensity and thus decreases in temperature at defect location [13]. Hence, the transducer must operate at its resonant frequency for effective ultrasonic stimulated thermography.

Furthermore, a modelling scheme was developed to simulate vibro-thermography of structures used in gas turbine engine components in order to reduce mass and enhanced cooling [14]. The framework of the model is comprised of coupled thermoplastic heat generation and various effects due to nonlinear vibration arising from excitation, engagement force on target structure by the ultrasonic horn, and structural boundary conditions. It has been observed from contact dynamic simulations that subharmonics, superharmonics, sub-superharmonic and chaos are all present in the structure even if the exciter is sharply tuned at the resonant frequency. This model is also used for obtaining spatio-temporal temperature distribution in the target structure that can be extended to understand vibro-thermographic characteristics in complex structures with hidden defects. Subsequently, a significant increase in SNR combined with suppression of effects due to non-uniform heating, background reflections and surface non-uniformities can be achieved by a novel adaptive spectral band integration procedure for post-processing of flash thermography data [15]. This procedure integrates spectral information of each individual pixel thus obtaining maximum detectability of defects such as flat bottom holes, BVIDs as well as a stiffened composite panel with production defects. The technique was also able to estimate the accurate size and depth of the defects when compared with pulsed phase thermography. Moreover, the same group of researchers compared the efficiency of time and frequency domain analysis techniques in flash thermography to improve the detectability of defects [16]. Single bin procedures such as thermal signal area and dynamic thermal tomography; and integrated bin procedures such as frequency domain tomography and adaptive spectral band integration are considered for the study. The single bin approach in the frequency domain showed better detectability of defect and higher defect depth estimation as compared to its time domain counterpart. Further, the integrated frequency domain approach shows higher sensitivity to non-uniform heating and the best defect detectability among all.

Subsequently, nonlinear acoustic wave spectroscopy (NAWS) has recently proved to be an efficient tool for detection of small defects in structures [17]. NAWS has an advantage over conventional non-destructive testing (NDT) techniques where the information about reflection of incident vibration waves play a major role in in detection of damages [18]. The interaction between defect and the incident wave leads to generation of higher harmonics which can be better understood using NAWS [19]. Local defect resonance (LDR) phenomenon is one of such NAWS technique that is emerging in recent years. Since the scattering and reflection data of the incident signal in case of smaller defects is not so prominent, NAWS and vibro-thermography technique are combined for detection of defects [20]. LDR based vibro-thermography technique is much advantageous and less time consuming than other NDT techniques for detection and location of small defects even in intricate structures. Moreover, the LDR based frictional heating was introduced which illustrated the rubbing and clapping action at defect site induces an internal heating that can be detected easily using an IR camera [21]. It was observed that the amplitude of vibration is quite high at the defect location when excited with its fundamental defect frequency. Subsequently, an IR camera was used with cell phone attachment for performing lock-in thermography [22]. With the help of image processing, sub-surface defects and dental caries were successfully detected. Also, a LDR based thermal imaging experiment was carried out to characterize various defects using air-coupled ultrasound excitation [23]. Further, defects were successfully activated using sweep excitation for low energy vibro-thermography technique [24]. The post-processing included a Fourier-transform of the temperature response obtained from the defect in case of unknown LDR frequencies. The LDR concept was also used in combination of IR imaging for detecting minute defects using a nonlinear ultrasonic stimulated thermography (NUST) method [25]. It is found that by exciting the structure at a frequency yielding nonlinear response may lead to increase in the temperature rise at defect area. The same research group again used sweep excitation for activating the LDR frequency in order to obtain higher heat generation at a crack surface [20].

The various literatures discussed on vibro-thermography and LDR based vibro-thermography suggests that the technique is widely accepted for detection, location and sizing of defects in composite structures. The advantage of such a technique is mainly for detection of small and complicated defects such as delaminations and barely visible impact damages (BVIDs) in composite structures. Hence, this chapter will address the problem of defect detection in glass fibre reinforced polymer (GFRP) composite by implementing LDR based vibro-thermography technique.

2. Theory

2.1 Local defect resonance (LDR)

The local defect resonance (LDR) is a phenomenon that occurs mainly due to presence of a defect in any structure. Whenever, a structure consists of damage, the local stiffness and mass of the structure at defect location changes. This change leads to a new parameter known as the effective stiffness and effective mass of the structure over the defect area. According to Solodov et al. [21], when the plate with defect is excited with a frequency that matches with the frequency of the defect due to its effective stiffness and effective mass, resonance occurs. This phenomenon of resonance at the defect area is termed as local defect resonance. The LDR phenomenon leads to high amplitude of vibration at the defect location while the rest of the plate remains at negligible amplitudes. Exciting a plate with one of its LDR frequency will lead to clapping and rubbing action at the corresponding defect. This clapping and rubbing action will further generate a local heat over the damage area. The analytical relation for detection of LDR frequency is given by Solodov et al. [21], in case of a flat bottom hole, which can be found analogous to a case of delamination in composites.

Figure 1 shows a schematic of four layered composite with central delamination. The analytical relation for calculating LDR frequency is based on parameters such as diameter of delamination (a) and residual thickness (t). The effective stiffness (K_e) of the structure at defect location is given by [26]

$$K_e = \frac{64\pi E t^3}{a^2(1-\nu^2)} \tag{1}$$

Where E is modulus of elasticity and ν is Poisson's ratio. Subsequently, the effective mass of delamination is given as [26].

$$M_e = \frac{9}{20}\pi \rho t a^2 \tag{2}$$

Finally, the analytical expression for calculating the LDR frequency of any defect in form of delamination is obtained by substituting Eqs. (**1**) and (**2**) in natural frequency relation of the structure, and is expressed as [26]

$$f_{LDR} = \frac{6.4t}{a^2}\sqrt{\frac{E}{12\rho(1-\nu^2)}} \tag{3}$$

Where ρ is mass density of the material.

2.2 LDR based vibro-thermography

The excitation of a structure having delamination with the LDR frequency corresponding to its damage leads to a rise in temperature at the defect location, as discussed previously. The heat generated at the defect location due to clapping and rubbing phenomena at the defect interface propagates away from the defect layer towards the surface of the structure. These thermal signatures are captured by using IR imaging and the technique is called as LDR based vibro-thermography. In this chapter, the LDR based vibro-thermography will be discussed by carrying out a

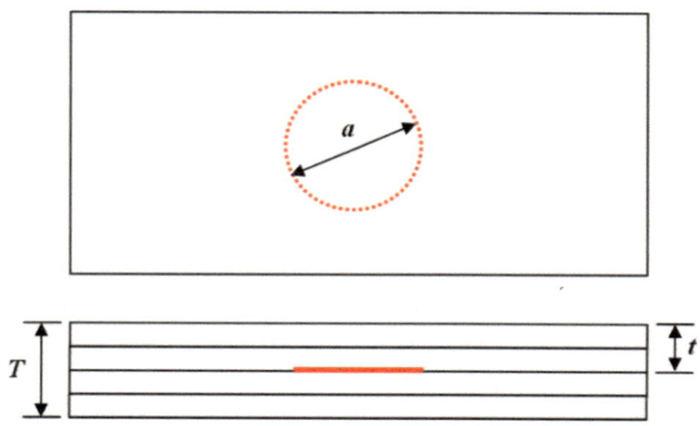

Figure 1.
Schematic of a four layered composite structure with circular delamination at the central layer.

explicit coupled temperature-displacement analysis. The nonlinear wave propagation analysis is carried out on glass fibre reinforced polymer (GFRP) composite in order to obtain the thermal response of the structure at the defect interface. The numerical model considered must fulfil the equilibrium condition in the deformed structure to obtain the solution of forces and displacements at specific nodes. The momentum equation based on principal of virtual displacement can be used for translation and rotational motion of the body under harmonic loading, and is given as [25].

$$\int_{\xi} \delta d\ \sigma d\xi + \int_{\xi} \delta v\ \rho \dot{v} d\xi + \int_{\xi} \delta v\ B d\xi - \int_{\varphi} \delta v\ S_t\ d\varphi = 0 \tag{4}$$

where ξ is current domain, φ is boundary of the body, δd is the virtual displacement, σ is stress, δv is virtual velocity, B is the body force vector and S_t is the surface traction. These terms represent the internal forces on the system, inertia force, body force and surface traction in the system, respectively.

The Courant-Friedrichs-Levy condition is satisfied while carrying out the explicit temperature-displacement analysis that requires integrating through small time increments [27]. The central-difference and forward-difference are required to be stable and hence a limit of time increment is chosen as

$$\Delta t \leq \min.\left(\frac{2}{\omega_{\max}}, \frac{2}{\lambda_{\max}}\right) \tag{5}$$

Where $\omega_{\max}$ is highest natural frequency of the structure and $\lambda_{\max}$ is largest eigenvalue for the solution. The time increment, Δt must be calculated such that the wavelength does not exceed more than a single element edge length. Therefore, the size of element and time increment during the temperature-displacement analysis is a critical factor that needs to be taken care of. Otherwise, higher time increment will lead to high amplitude oscillation of the time history variables and an unstable solution.

During the coupled temperature-displacement analysis, internal stress and strain are developed due to harmonic excitations from the incident wave which further results in temperature rise at the defect location. The transient heat transfer occurring at the defect surface is computed using the following equation [25].

$$\rho C_p \frac{\partial T}{\partial t} = \nabla \phi + H_g \tag{6}$$

Where C_p is specific heat capacity, T is time dependent temperature field, ϕ is heat flux vector per unit volume and H_g is total internal heat generated per unit volume. From the Fourier's heat conduction law, the time dependent temperature during each time step can be calculated which is discretised and expressed as [25].

$$T_{n+1} = \left(\frac{C_n}{\Delta t} + \gamma_n\right)^{-1} \left(F_n + T_n \frac{C_n}{\Delta t}\right) \tag{7}$$

Where n is the analysis step (n = 1, 2, 3, ...), C is heat capacity matrix, γ is conductivity matrix and F is thermal force vector. The temperature rise at the defect location due to internal heat when the plate is excited at LDR frequency can be evaluated using the above equation. Therefore, vibro-thermography can be performed on a structure by incorporating LDR excitations that lead to higher amplitude of vibration at defect area causing a temperature rise.

3. Numerical analysis

This section discusses the numerical simulation carried out for detection of sub-surface delamination in a four layered GFRP plate with circular delamination at the central layer. The numerical model of the delaminated GFRP plate was carried out using ABAQUS 6.14 software. Each layer of the GFRP plate is considered to be of 0.5 mm thickness with the total thickness of plate as 2 mm. The area of the plate considered for numerical simulation is 180 × 180 mm^2 with a delamination between second and third layer. Thus, the residual thickness of the damage required for calculation of analytical LDR frequency is 1 mm. The diameter of the circular delamination is taken as 12 mm while the location of the delamination is at x = 120 mm and y = 60 mm, taking left bottom corner of the plate as the origin. **Figure 2** shows the schematic of the delaminated plate with circular delamination.

The material properties of GFRP plate used for carrying out the explicit temperature displacement analysis are given in **Table 1**. The rule of mixtures was implemented for calculating the material properties of GFRP specimen from the individual property of glass fibre and epoxy [29]. The simulation is carried out by providing sliding contact interaction properties at the defect interface. Moreover, the friction properties are included on the delamination area while the rest of the plate area is provided with tie constraints for all layers. The initial

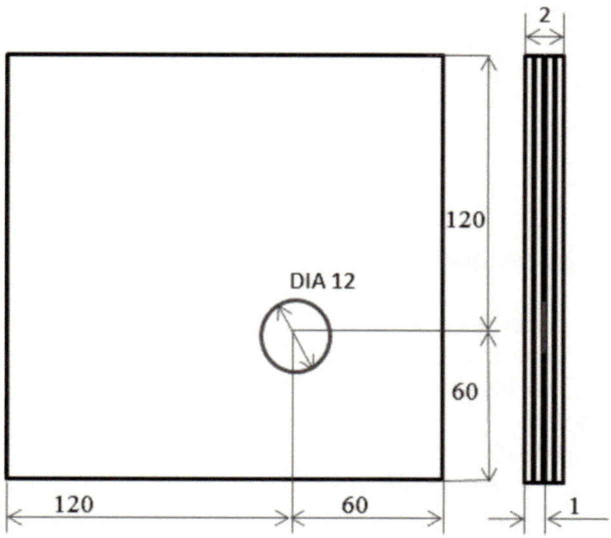

Figure 2.
Schematic of GFRP plate with circular delamination at the central layer.

Material	Density (in kg/m^3)	Young's Modulus (in GPa)	Poisson's ratio	Specific heat (J/kg K)	Thermal conductivity (W/m K)
Epoxy	1200	3.30	0.35	1100.0	0.17
E-glass fibre	2540	28.0	0.22	800.81	1.35
GFRP plate	2145	10.7	0.30	908.25	0.94

Table 1.
Material properties for GFRP plate [28].

ambient temperature was considered as 293 K for all nodes of the plate. The number of elements considered for carrying out the present numerical investigation is 42,269 10-noded tetrahedral elements. First, the analytical LDR frequency of the delamination is calculated from the relation presented in previous section, followed by the steady state dynamic analysis to confirm the LDR frequency. The steady state analysis is performed on a frequency limit such that the analytical LDR calculated falls within the range of the analysis. The mode shape obtained from the contour plot of steady state analysis confirms the exact LDR frequency required for performing explicit temperature-displacement analysis. Subsequently, the explicit coupled temperature-displacement analysis is carried out followed by heat transfer analysis to detect the defect using LDR based vibro-thermography.

The explicit coupled temperature-displacement analysis is performed on the GFRP plate using single periodic LDR excitation confirmed from the steady state analysis. The force amplitude for carrying out the analysis is taken as 50 N in the thickness direction. The excitation time of 10 ms is used for vibrating the plate at its LDR frequency with a fixed time increment of 5×10^{-8} s. The time increment is calculated according to the CFL condition for explicit analysis [27]. The 10-noded modified thermally coupled second order tetrahedron (C3D10MT) mesh elements are used for the analysis. Subsequently, a node on the top surface of the plate positioned at centre of the damage area is selected to determine the amplitude of vibration during excitation. In the analysis history output, nodes at delamination layer are selected for plotting temperature profiles. The temperature gradient obtained at the end of explicit coupled temperature-displacement analysis is further used as predefined field for a heat transfer analysis. Finally, a 10 ms heat transfer analysis is performed with 10-noded quadratic heat transfer tetrahedron (DC3D10) mesh type for this step. The transient response of the heat transfer step is captured at the delamination layer as well as the top surface of the plate.

4. Results and discussions

This section discusses the results obtained from explicit coupled temperature-displacement analysis followed by heat transfer analysis in case of LDR frequency excitation based vibro-thermography. First, a steady state analysis of the GFRP plate is performed with a frequency range around the analytical LDR frequency calculated using Eq. (3). The mode shape confirming the LDR frequency of the delamination is found to be 24.8 kHz from the steady state analysis, as shown in **Figure 3**. The LDR frequency excitation in form of single periodic continuous excitation is applied on the transmitter nodes of the numerical model from the top surface.

On performing the explicit coupled temperature-displacement analysis, the amplitude-frequency spectra obtained from a node selected on the top surface of the plate above the delamination location is presented in **Figure 4**. The highest peak in the frequency spectra is obtained at 24.86 kHz which is close to the result obtained from steady state analysis. The highest amplitude of 6.488×10^{-6} m was observed at the delamination area.

Subsequently, the average nodal temperature profile at the delamination location during the span of explicit dynamic coupled temperature-displacement analysis is plotted in **Figure 5**. The plot shows variation of temperature at the delamination interface with respect to time of excitation. It is observed that the average nodal temperature at delamination location starts rising after 4 ms of excitation. The nodal temperature at delamination region rises gradually to around

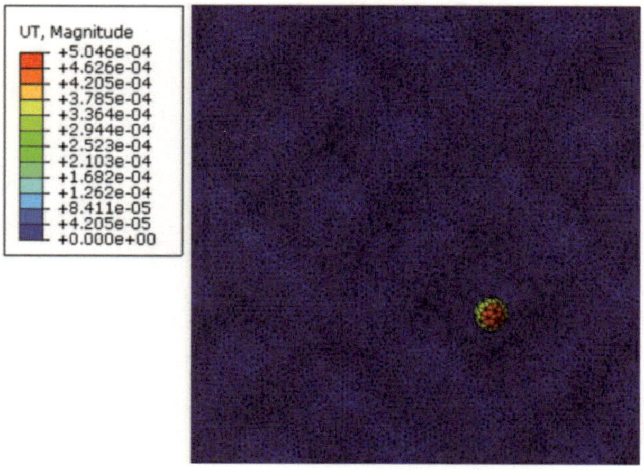

Figure 3.
Contour plot of the GFRP plate depicting the mode shape of the delamination at LDR frequency of 24.8 kHz.

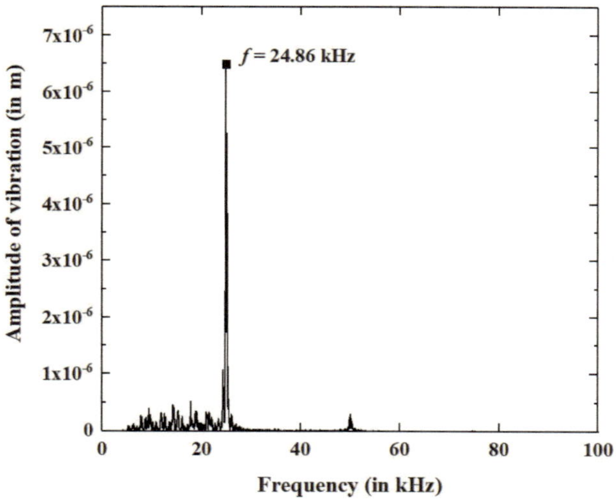

Figure 4.
Frequency spectra obtained at the delamination location on the top surface of the plate for single periodic LDR excitation.

293.015 K in 10 ms. Since, the temperature rise is not saturated at the end of excitation, the plate can be excited for more time until the temperature rise saturates.

Moreover, all the elements over the delamination area is selected to plot the elemental temperature rise over delamination during 10 ms of heat transfer analysis, as illustrated in **Figure 6**. The plot shows a rise of temperature from the ambient for some of the elements over the delamination area. It is observed that the maximum temperature recorded during heat transfer analysis is 293.15 K. Since the temperature rise over the delamination region is around 150 mK, the vibrothermography scanning can be done at this instance of time with any IR camera that normally has a resolution of 30 mK.

Furthermore, contour plot for nodal temperature on the delamination surface of GFRP plate after explicit temperature-displacement analysis is illustrated in **Figure 7**. It is clear from the contour plot that the temperature rise is observed only

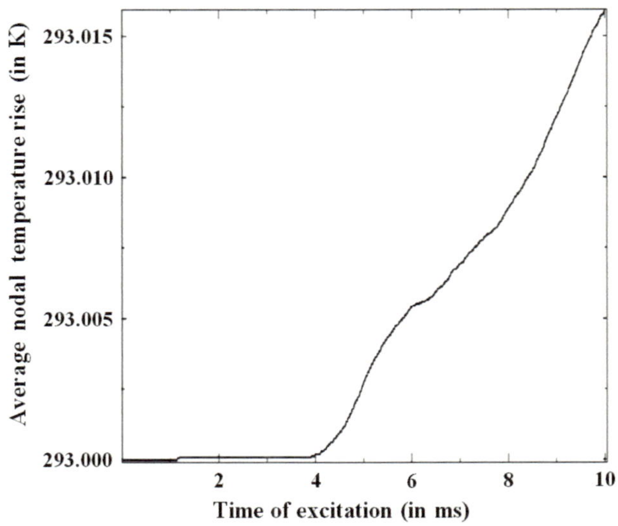

Figure 5.
Average nodal temperature rise at delamination layer for single periodic continuous LDR excitation.

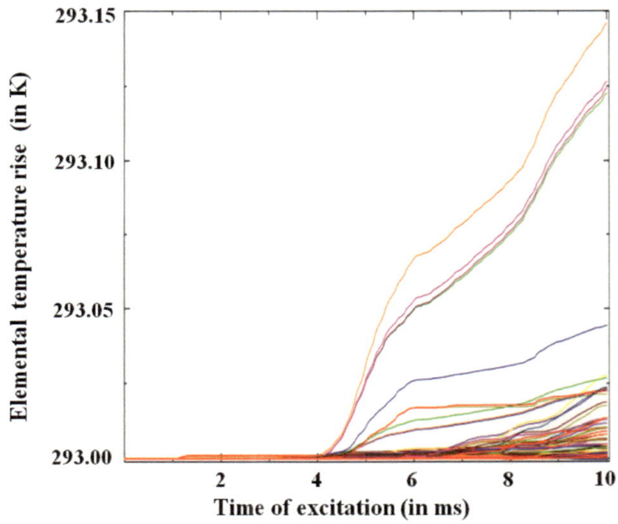

Figure 6.
Elemental temperature rise at delamination layer for single periodic continuous LDR excitation.

at the delamination location while the rest of the plate shows no thermal signatures at all. However, it is clear that the whole delamination area is not covered by the heat flow at this time instance. Hence, the excitation of the plate must be done for more time period such that the whole delamination area is covered by the heat distribution. Similarly, the elemental temperature is shown in contour plot depicted in **Figure 8**, where heat is distributed to a larger area after heat transfer analysis as compared to the explicit temperature-displacement analysis. Although, the whole delamination is not covered at this instance and more time for excitation is required. The defect imaging at later time increment is advantageous because of better judgement and idea of defect shape and size is detected easily. This is one of the merits of performing LDR based vibro-thermography technique that provides high contrast defect images that can be captured using an IR camera.

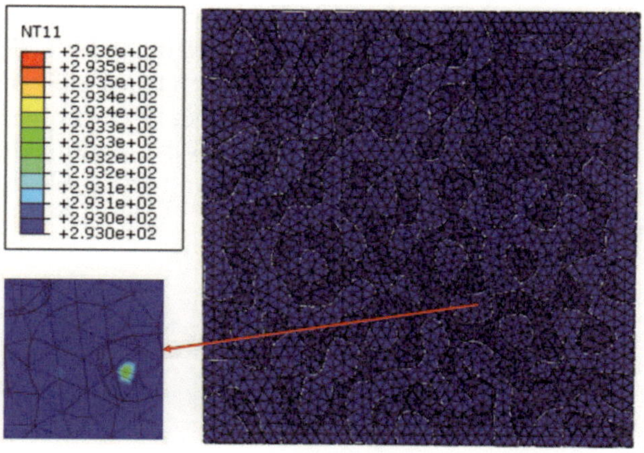

Figure 7.
Contour plot of nodal temperature on the delamination surface of delaminated GFRP plate after heat transfer analysis.

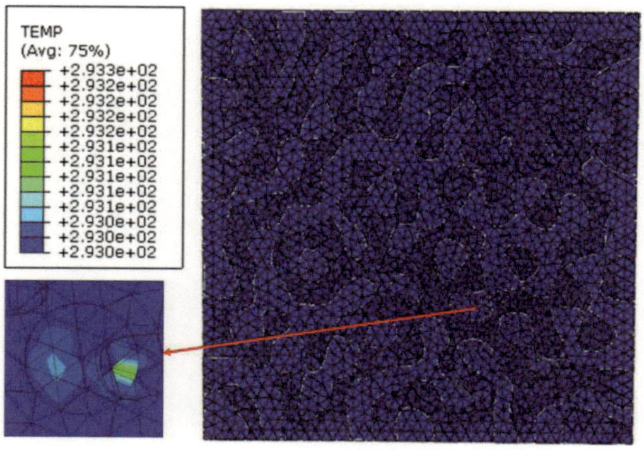

Figure 8.
Contour plot of nodal temperature on the top surface of delaminated GFRP plate at 10 ms during single periodic continuous LDR excitations.

5. Conclusions

In this chapter, a LDR frequency excitation based vibro-thermography technique is implemented for detection of a circular delamination in a four layered GFRP plate. The analytical LDR frequency is calculated using the expression discussed in the theory section. The analytical LDR frequency is then confirmed by carrying out a steady state dynamic analysis. The mode shape obtained from the contour plot of the steady state analysis confirms the LDR frequency of the delamination which is further used for carrying out the explicit dynamic coupled temperature-displacement analysis. Subsequently, the amplitude-frequency spectra during the temperature-displacement analysis is plotted to observe a rise in temperature at the defect location due to clapping and rubbing action during resonance. The average nodal at delamination layer is presented to find the maximum temperature rise at delamination region when the plate is excited with LDR frequency. Moreover, a 10 ms heat transfer analysis is carried out to obtain high contrast temperature response on top surface of the GFRP plate. Finally, the temperature profile is plotted for all elements over the delamination layer after heat transfer analysis. The temperature profile plotted at the defect interface signifies that the temperature rises drastically when the structure is excited with LDR frequency and subsequently, enhances the chance of defect detection using vibro-thermography technique. The vibro-thermographic imaging using an IR camera is recommended after longer period of LDR excitation. Thus, the LDR based vibro-thermography is found to be an efficient tool for detection, location and size estimation of defects in composite structures.

Author details

Subhankar Roy and Tanmoy Bose*
Department of Mechanical Engineering, National Institute of Technology Meghalaya, Shillong, Meghalaya, India

*Address all correspondence to: tanmoy.jgec04@gmail.com

References

[1] Henneke EG, Reifsnider KL, Stinchcomb WW. Thermography—An NDI method for damage detection. The Journal of the Minerals, Metals & Materials Society. 1979;**31**(9):11-15

[2] Pye CJ, Adams RD. Detection of damage in fibre reinforced plastics using thermal fields generated during resonant vibration. NDT International. 1981;**14**(3):111-118

[3] Rantala J, Wu D, Busse G. Amplitude-modulated lock-in vibrothermography for NDE of polymers and composites. Research in Nondestructive Evaluation. 1996;**7**: 215-228

[4] Wu D, Busse G. Lock-in thermography for nondestructive evaluation of materials. Revue Générale de Thermique. 1998;**37**(8):693-703

[5] Han X, Zeng Z, Li W, Islam Md S, Lu J, Loggins V, et al. Acoustic chaos for enhanced detectability of cracks by sonic infrared imaging. Journal of Applied Physics. 2004;**95**(7):3792-3797

[6] Renshaw J, Holland SD, Thompson RB. Measurement of crack opening stresses and crack closure stress profiles from heat generation in vibrating cracks. Applied Physics Letters. 2008;**93**(8):081914

[7] Junyan L, Qingju T, Yang W. The study of inspection on SiC coated carbon–carbon composite with subsurface defects by lock-in thermography. Composites Science and Technology. 2012;**72**(11): 1240-1250

[8] Mendioroz A, Castelo A, Celorrio R, Salazar A. Characterization of vertical buried defects using lock-in vibrothermography: I. Direct problem. Measurement Science and Technology. 2013;**24**(6):065601

[9] Castelo A, Mendioroz A, Celorrio R, Salazar A. Optimizing the inversion protocol to determine the geometry of vertical cracks from lock-in vibrothermography. Journal of Nondestructive Evaluation. 2017; **36**(1):3

[10] Ciampa F, Mahmoodi P, Pinto F, Meo M. Recent advances in active infrared thermography for non-destructive testing of aerospace components. Sensors. 2018;**18**(2):609

[11] He Y, Chen S, Zhou D, Huang S, Wang P. Shared excitation based nonlinear ultrasound and vibrothermography testing for CFRP barely visible impact damage inspection. IEEE Transactions on Industrial Informatics. 2018;**14**(12): 5575-5584

[12] Derusova DA, Vavilov VP, Guo X, Druzhinin NV. Comparing the efficiency of ultrasonic infrared thermography under high-power and resonant stimulation of impact damage in a CFRP composite. Russian Journal of Nondestructive Testing. 2018;**54**(5): 356-362

[13] Derusova DA, Vavilov VP, Guo X, Shpil'noi VY, Danilin NS. Infrared thermographic testing of hybrid materials using high-power ultrasonic stimulation. Russian Journal of Nondestructive Testing. 2018;**54**(10): 733-739

[14] Geetha GK, Mahapatra DR. Modeling and simulation of vibro-thermography including nonlinear contact dynamics of ultrasonic actuator. Ultrasonics. 2019;**93**:81-92

[15] Poelman G, Hedayatrasa S, Segers J, Van Paepegem W, Kersemans M. Adaptive spectral band integration in flash thermography: Enhanced defect detectability and quantification in

composites. Composites Part B: Engineering. 2020;**202**:108305

[16] Poelman G, Hedayatrasa S, Segers J, Van Paepegem W, Kersemans M. An experimental study on the defect detectability of time- and frequency-domain analyses for flash thermography. Applied Sciences. 2020; **10**(22):8051

[17] Solodov I, Bai J, Busse G. Resonant ultrasound spectroscopy of defects: Case study of flat-bottomed holes. Journal of Applied Physics. 2013;**113**(22):223512

[18] Lima WJND, Hamilton MF. Finite-amplitude waves in isotropic elastic plates. Journal of Sound and Vibration. 2003;**265**:819-839

[19] Ciampa F, Pickering SG, Scarselli G, Meo M. Nonlinear imaging of damage in composite structures using sparse ultrasonic sensor arrays: Nonlinear imaging of damage. Structural Control and Health Monitoring. 2017;**24**(5): e1911

[20] Fierro GPM, Ginzburg D, Ciampa F, Meo M. Imaging of barely visible impact damage on a complex composite stiffened panel using a nonlinear ultrasound stimulated thermography approach. Journal of Nondestructive Evaluation. 2017;**36**(4):69

[21] Solodov I, Bai J, Bekgulyan S, Busse G. A local defect resonance to enhance acoustic wave-defect interaction in ultrasonic nondestructive evaluation. Applied Physics Letters. 2011;**99**(21):211911

[22] Razani M, Parkhimchyk A, Tabatabaei N. Lock-in thermography using a cell phone attachment infrared camera. AIP Advances. 2018;**8**:035305

[23] Solodov I, Derusova D, Rahammer M. Thermosonic Chladni figures for defect-selective imaging. Ultrasonics. 2015;**60**:1-5

[24] Rahammer M, Kreutzbruck M. Fourier-transform vibrothermography with frequency sweep excitation utilizing local defect resonances. NDT & E International. 2017;**86**:83-88

[25] Fierro GPM, Calla D, Ginzburg D, Ciampa F, Meo M. Nonlinear ultrasonic stimulated thermography for damage assessment in isotropic fatigued structures. Journal of Sound and Vibration. 2017;**404**:102-115

[26] Solodo I, Rahammer M, Gulnizkij N, Kreutzbruck M. Noncontact sonic NDE and defect imaging via local defect resonance. Journal of Nondestructive Evaluation. 2016;**35**(3):48

[27] Courant R, Friedrichs K, Lewy H. On the partial difference equations of mathematical physics. IBM Journal of Research and Development. 1967;**11**(2): 215-234

[28] AZo Materials—Online Publication for Materials Science Community [Internet]. Available from: https://www.azom.com/properties.aspx?ArticleID=764 [Accessed: 11 October 2021]

[29] Liu GR. A step-by-step method of rule-of-mixture of fiber- and particle-reinforced composite materials. Composite Structures. 1998;**40**(3–4): 313-322

Chapter 5

Periodic Acoustic Black Holes to Mitigate Sound Radiation from Cylindrical Structures

Jie Deng and Nansha Gao

Abstract

Acoustic black holes (ABHs) achieved by progressively diminishing structural thickness have been proved a very efficient approach for wideband vibration reduction, sound suppression, energy harvesting, and wave manipulation. In this chapter, the focus is placed on mitigating the sound emitted from cylindrical shells with embedded ABHs. In the applications of aeronautics, astronautics, and underwater vehicles, cylindrical shells are very common yet the vibroacoustic problems in such structures are very challenging. Even the researches on ABHs for straight beams and flat plates are boosting in recent years, the ABH effect is unclear for curved structures thus it deserves further investigations to push forward their applications. Since cylindrical shells are usually long in, for example, airplanes and rockets, periodic ABHs are designed to alleviate the acoustic emission from them. The Gaussian expansion method (GEM) is employed to recover the vibration field on the cylinder and, based on that, the sound radiation model is developed to determine the emitted sound power level (SWL). The band gaps (BGs) are shown for infinite periodic ABH shells, followed by the vibroacoustic level for a finite periodic shell. Particularly, axial stiffeners are introduced and the influences of their quantity and width are carried out.

Keywords: acoustic black holes, acoustic radiation, cylindrical shells, band gaps, stiffeners

1. Introduction

An acoustic black hole (ABH) is usually realized by reducing structural thickness following a power law $h(x) = \varepsilon x^m \; (m \geq 2)$, as illustrated in **Figure 1**. When an incident flexural wave impinges at the edge of the ABH, its wavelength and wave speed get gradually decreased. Meanwhile, the wavenumber is however increased and the amplitude is intensified. In the ideal case, the thickness at the ABH tip decays to zero, where the wave velocity vanishes as well, such that the traveling time to its center becomes infinite. In other words, the wave will never reach the tip. In analogy with cosmology, the termination behaves like a "Black Hole" in which nothing can escape from it. This is the story of how the term "Acoustic Black Hole" was coined [1]. Howbeit, in real applications, generally, an ABH is imperfect. Namely, there exists a truncation near the ABH tip, which results in obvious reflection because of the residual thickness [2]. Fortunately, attaching a thin viscoelastic

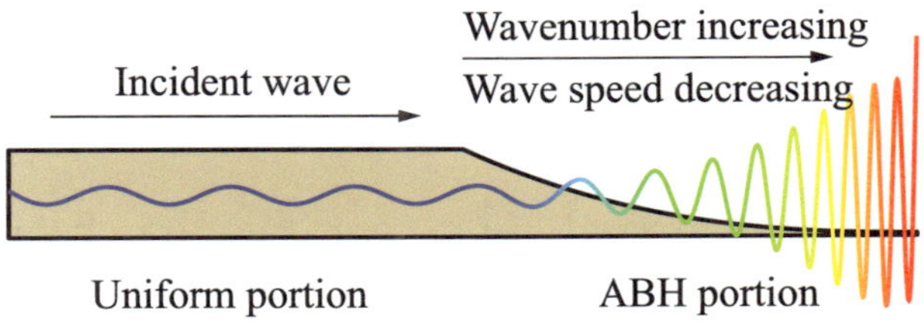

Figure 1.
Illustration of ABH effect: the incident wave is localized in the ABH tip as it propagates toward the ABH.

layer at the ABH tip, where the energy is highly concentrated, can alleviate this problem [3]. Recently, constrained viscoelastic layers have been suggested to enhance the damping effects, by changing the normal tensile and compressive deformation of damping material into the shear one [4].

It has been shown that the ABHs are very efficient to reduce vibration from straight beams [5, 6] and flat plates [7, 8]. The shapes and lengths of the damping layers have been extensively investigated [4, 9]. Also, different ABH designs have been proposed for the purpose of enhancing energy consumption [7, 10–12]. Thanks to the vibration reduction because of highly efficient damping, the sound radiation from ABH structures is accordingly reduced [13]. Not only that, recent studies have shown that the ABHs can also impair the sound radiation efficiency because of the thickness reduction [14, 15]. Particularly, for cavity noise, the ABH profile can destroy the coupling strength between structural and acoustic modes, which is the third underlying mechanism of the ABHs for reducing room noise [16, 17]. Furthermore, periodic and gradient-index ABHs are investigated for steering waves [18–20]. It is also worthwhile mentioning that ABHs can also enhance energy harvesting due to wave focalization [21], using piezoelectric layers rather than viscoelastic ones.

The state-of-the-art reviewed above are mainly centered on flat structures. However, in aeronautics, astronautics, and underwater vehicles, cylindrical shells are very common. There, the vibroacoustic problems are very critical to determine their comfortability and safety, thus it is very demanding to apply ABH features on them. Our previous efforts have been focused on the vibration of cylindrical beams [22] and shells [23, 24], together with the sound radiation from a finite cylinder [25]. In this chapter, we continue this topic but analyze the sound radiation from periodic ABH shells.

As shown in **Figure 2a**, an infinite periodic ABH shell is considered, with each unit cell having radius R and length L_{cell}. An ABH plus a thin viscoelastic layer (see the green layer) is laid in the center of the cell. The geometries of the ABH and the damping layer are detailed in **Figure 2b**. Here, the profile of the ABH is defined by $h(x) = \varepsilon|x|^m + h_c$, where $\varepsilon = (h_{uni} - h_c)r_{abh}^{-m}$ stands for the ABH slope and, r_{abh}, h_c, and m respectively are the ABH radius, residual thickness, and order. We will characterize the band gaps (BGs) for infinite periodic ABH shells and their dependence on the ABH geometry. Next, a finite periodic ABH shell containing five cells will be characterized, under a ring excitation acting at x_f (see **Figure 2c**). The translational springs $k_i, i = 1, 2$ and rotational ones $p_i, i = 1, 2$ are intended for boundary conditions (distributed circumferentially). As one could expect, the appearance of ABHs weakens the stiffness of the whole structure. This may deteriorate the structural problems. To partially solve this, we can introduce N stiffeners for each cell (see **Figure 2d**), with each width being W. The effects of the stiffeners will also be investigated at the end of this chapter.

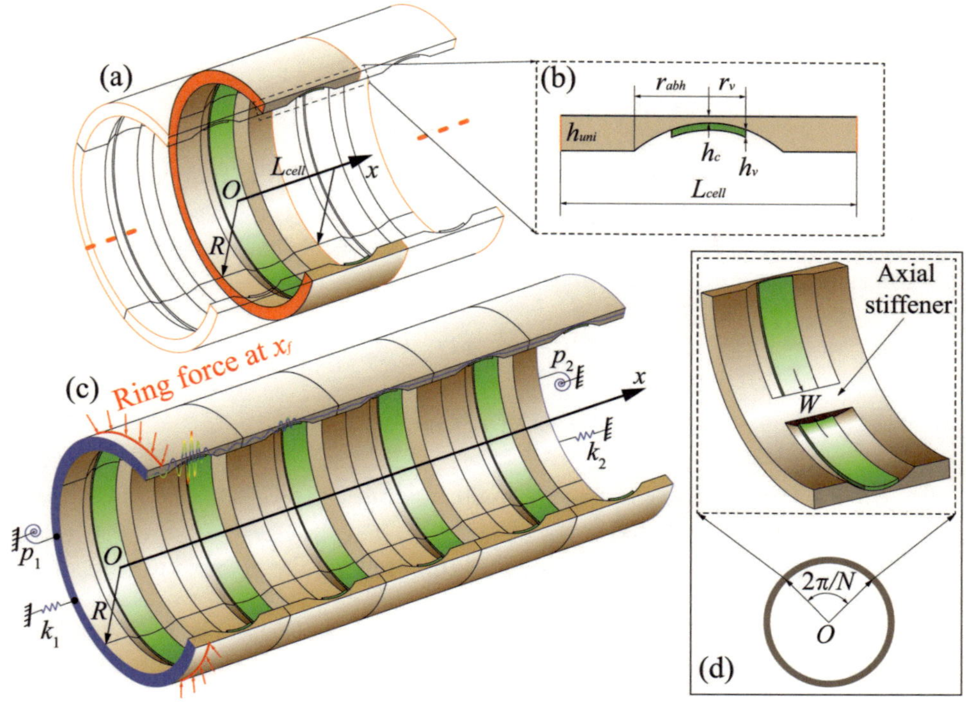

Figure 2.
Illustration of the periodic ABH cylindrical shell. (a) A unit ABH cell having a thin damping layer (green). (b) The geometrical detail of the ABH profile. (c) A supported finite ABH cell having five cells under a ring excitation. (d) Illustration of the stiffener for enhancing the structure (with N stiffeners in the circumferential direction).

2. Vibration characteristics of infinite periodic ABH shells

2.1 Gaussian expansion for the vibration of infinite periodic ABH shells

The goal of this section is that of developing a semi-analytical model for characterizing the vibration of the infinite periodic ABH shell. Let us consider three variables, u, v, and w, which are the displacements in the axial, circumferential, and radial directions, respectively. They can be decomposed by

$$u(x,y,t) = \sum_i a_i(t)\psi_i(x,y) = \boldsymbol{a}^\top\boldsymbol{\psi} = \boldsymbol{\psi}^\top\boldsymbol{a}, \tag{1}$$

$$v(x,y,t) = \sum_i b_i(t)\xi_i(x,y) = \boldsymbol{b}^\top\boldsymbol{\xi} = \boldsymbol{\xi}^\top\boldsymbol{b}, \tag{2}$$

$$w(x,y,t) = \sum_i c_i(t)\varphi_i(x,y) = \boldsymbol{c}^\top\boldsymbol{\varphi} = \boldsymbol{\varphi}^\top\boldsymbol{c}, \tag{3}$$

where

$$\boldsymbol{a} = \hat{\boldsymbol{A}}\exp(\mathrm{j}\omega t), \boldsymbol{b} = \hat{\boldsymbol{B}}\exp(\mathrm{j}\omega t), \boldsymbol{c} = \hat{\boldsymbol{C}}\exp(\mathrm{j}\omega t), \tag{4}$$

are the coefficient vectors to be determined, while $\boldsymbol{\psi}, \boldsymbol{\xi}$, and $\boldsymbol{\varphi}$ are the shape function vectors with entries $\psi_i(x,y), \xi_i(x,y)$, and $\varphi_i(x,y)$, respectively. With the aid of Kronecker product, the vectors $\boldsymbol{\psi}, \boldsymbol{\xi}$, and $\boldsymbol{\varphi}$ can be factorized as

$$\boldsymbol{\psi}(x,y) = \boldsymbol{\alpha}^{\psi}(x) \otimes \boldsymbol{\beta}^{\psi}(y), \tag{5}$$

$$\boldsymbol{\xi}(x,y) = \boldsymbol{\alpha}^{\xi}(x) \otimes \boldsymbol{\beta}^{\xi}(y), \tag{6}$$

$$\boldsymbol{\varphi}(x,y) = \boldsymbol{\alpha}^{\varphi}(x) \otimes \boldsymbol{\beta}^{\varphi}(y), \tag{7}$$

where $\boldsymbol{\alpha}^{(i)}(i = \psi, \xi, \varphi)$ are column vectors containing basis functions depending on the x direction yet $\boldsymbol{\beta}^{(i)}(i = \psi, \xi, \varphi)$ the ones on the y direction. To accurately capture the localized displacements in the ABH portion, the entries of $\boldsymbol{\alpha}$ and $\boldsymbol{\beta}$ are selected as Gaussian functions

$$\alpha_i(x) = 2^{s_x/2} \exp\left[-(2^{s_x}x - q_{xi})^2/2\right], \tag{8}$$

$$\beta_i(y) = 2^{s_y/2} \exp\left[-\left(2^{s_y}y - q_{yi}\right)^2/2\right], \tag{9}$$

in which s_x and s_y are the scaling parameters, while q_x and q_y are the translational parameters, in the x and y directions, respectively. For brevity, readers are referred to our previous works [4, 26] to thoroughly comprehend the detailed process of how to produce acceptable basis.

Unlike finite and flat plates, bear in mind that the displacements, on the one hand, must be continuous in the circumferential direction ($\mathbb{C}^0$ and $\mathbb{C}^1$ are enough, see [22]), which requires

$$\mathbb{C}^0 : u(x, -\pi R) = u(x, \pi R), \;\; v(x, -\pi R) = v(x, \pi R), \;\; w(x, -\pi R) = w(x, \pi R), \tag{10}$$

$$\mathbb{C}^1 : \frac{\partial u}{\partial y}(x, -\pi R) = \frac{\partial u}{\partial y}(x, \pi R), \;\; \frac{\partial v}{\partial y}(x, -\pi R) = \frac{\partial v}{\partial y}(x, \pi R), \;\; \frac{\partial w}{\partial y}(x, -\pi R) = \frac{\partial w}{\partial y}(x, \pi R). \tag{11}$$

On the other hand, Bloch-Floquet periodic boundary conditions must be imposed in the axial direction for a unit cell ($\mathbb{C}^0$ and $\mathbb{C}^1$, too), namely

$$\mathbb{C}^0 : u(0,y) = u(L_{cell},y)\lambda, \;\; v(0,y) = v(L_{cell},y)\lambda, \;\; w(0,y) = w(L_{cell},y)\lambda, \tag{12}$$

$$\mathbb{C}^1 : \frac{\partial u}{\partial x}(0,y) = \frac{\partial u}{\partial x}(L_{cell},y)\lambda, \;\; \frac{\partial v}{\partial x}(0,y) = \frac{\partial v}{\partial x}(L_{cell},y)\lambda, \;\; \frac{\partial w}{\partial x}(0,y) = \frac{\partial w}{\partial x}(L_{cell},y)\lambda, \tag{13}$$

where $\lambda = \exp\left(\mathrm{j}k_x L_{cell}\right)$, k_x is the axial wavenumber in the irreducible Bernouin zone [18]. Via implementing the reconstruction process in [18], the continuity in the circumferential direction (Eqs. (10) and (11)) and the periodicity in the axial one (Eqs. (12) and (13)) can be satisfied.

Provided the kinetic energy, K, and the potential one, U, are presented in terms of u, v, and w [27], the Lagrangian of the whole system can be built

$$L = K - U = \frac{1}{2}\dot{\boldsymbol{q}}^{\top}\boldsymbol{M}\dot{\boldsymbol{q}} - \frac{1}{2}\boldsymbol{q}^{\top}\boldsymbol{K}\boldsymbol{q}, \tag{14}$$

where

$$\boldsymbol{q} = \left[\hat{\boldsymbol{A}}^{\top}, \hat{\boldsymbol{B}}^{\top}, \hat{\boldsymbol{C}}^{\top}\right]^{\top} \exp\left(\mathrm{j}\omega t\right) \equiv \hat{\boldsymbol{Q}} \exp\left(\mathrm{j}\omega t\right), \tag{15}$$

represents the assembled undetermined time-dependent vector related to admissible shape functions. $\boldsymbol{M}$ represents the mass matrix and $\boldsymbol{K}$ the stiffness one.

Finally, applying the Euler–Lagrange equations $\partial_t(\partial_{\dot{q}}L) - \partial_q L = \mathbf{0}$ to Eq. (14) yields the equations of motion in the frequency domain,

$$(-\omega^2 \boldsymbol{M} + \boldsymbol{K})\hat{\boldsymbol{Q}} = \mathbf{0}, \tag{16}$$

whose solution permits calculating the dispersion curves and eigenmodes for infinite periodic ABH shells.

2.2 Numerical results

2.2.1 Dispersion curves and band gaps

The dispersion curves of an ABH cell, whose geometry and material are detailed in **Table 1**, have been carried out and plotted in **Figure 3a**. For the purpose of validation, the result from a reference FEM model has also been included in **Figure 3a**. From the figure, it is seen that the two results are very close at each

Geometry parameters	Material parameters
$m = 2.8$	$\rho = 7800\ \text{kg/m}^3$
$L_{cell} = 1\ \text{m}$	$E = 210\ \text{GPa}$
$R = 5\ \text{m}$	$\eta = 0.005$
$h_{uni} = 0.03\ \text{m}$	$\nu = 0.3$
$r_{abh} = 0.25\ \text{m}$	
$\varepsilon = 1.3581\ \text{m}^{-1.8}$	$\rho_v = 950\ \text{kg/m}^3$
$h_c = 0.002\ \text{m}$	$E_v = 5\ \text{GPa}$
$r_v = 0.125\ \text{m}$	$\eta_v = 0.5$
$h_v = 0.008\ \text{m}$	$\nu_v = 0.3$

ρ, shell density; ρ_v, damping layer density; E, shell young modulus; E_v, damping layer young modulus; η, shell loss factor; η_v, damping layer loss factor; ν, shell Poisson ratio; ν_v, damping layer Poisson ratio.

Table 1.
Geometry and material parameters of the ABH cylindrical shell.

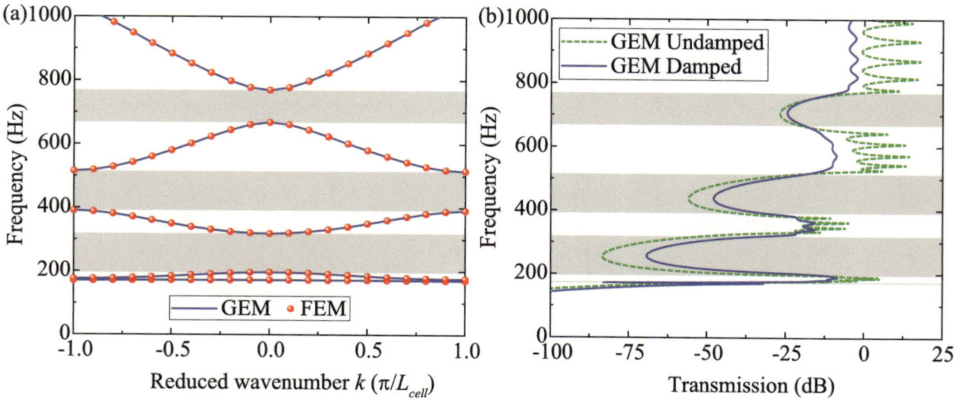

Figure 3.
(a) Dispersion curves together with band gaps calculated with GEM and FEM. (b) Transmission from the left end to the right for a finite shell having five cells with and without damping layers, carried out with the GEM.

wavenumber, indicating the correctness of the present GEM model. Most importantly, four-band gaps (BGs) are observed within 1000 Hz, in which the first one is very small. Using the GEM without axial periodic boundary conditions for a finite shell having five cells, we can compute the transmission from one end to the other. As shown in **Figure 3b**, the transmission is very low at BG frequencies. For the shell without damping layers (undamped), the transmission is strong in the passbands. However, this situation can be ameliorated after implementing the damping layers, with a maximum reduction of up to ~25 dB.

To reveal the mechanism of the BGs, we have computed the first six eigenmodes of the unit cell, at wavenumber $k_x = 0$ for λ. As illustrated in **Figure 4**, for most orders, the vibration is very strong in the ABH area. While this is not the case for the 1-st order (see **Figure 4a**). This is because of the ring frequency (173 Hz), below which the cylindrical shell is almost not vibrating in the radial direction. However, for the 2-nd to 6-th orders (see **Figure 4b–f**), the wave is gradually concentrating in the ABH portion, belonging to the locally resonant effect. That is, the BGs shown in **Figure 3** are locally resonant ones, similar to the periodic ABH beams reported in [28].

2.2.2 Parametric analysis: effects of the ABH order, central thickness, and radius

The ABH profile is generally controlled by three parameters, m, h_c, and r_{abh}, which represent the ABH order, central thickness, and radius, respectively. It is worthwhile testing how these parameters affect the BGs.

Let us first look at the influence of the ABH order. As shown in **Figure 5a**, the 2-nd BG starts to gradually decrease as m increases. The 3-rd BG is however distinctive because the BG inverses near $m = 3$, then the upper bound almost keeps still but its lower bound drops. For the 4-th one, the width of the BG first becomes larger then turns out smaller as m goes up. In general, the total width of the four BGs almost remains the same, but they will be more compact and converge to lower frequencies. Note that we have also included the changes of the BGs for the ABH shell with a damping layer (damped). It is seen that the phenomenon is close to the

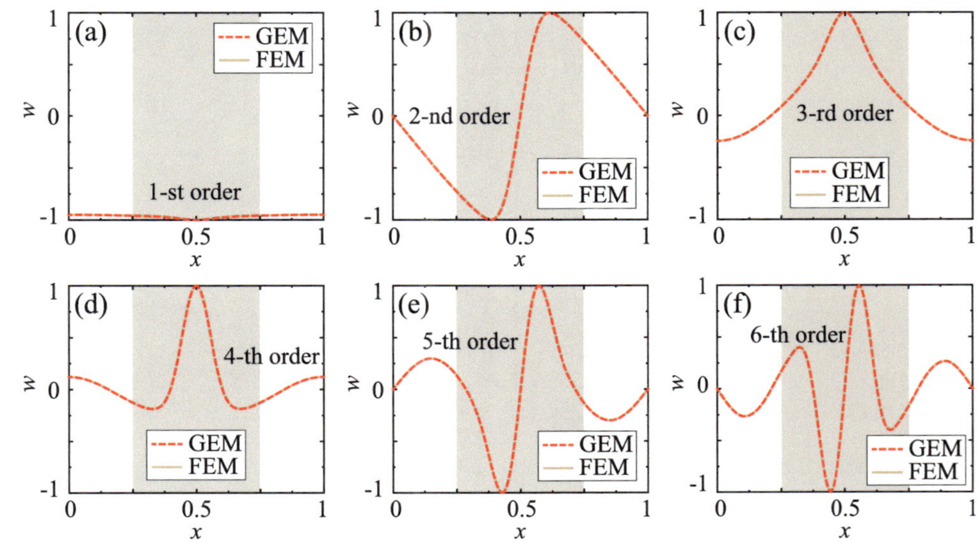

Figure 4.
The first 6 eigenmodes for infinite periodic cells at wavenumber $k_x = 0$. The shaded area stands for the ABH portion. (a)-(f) respectively correspond to the 1-st to the 6-th orders.

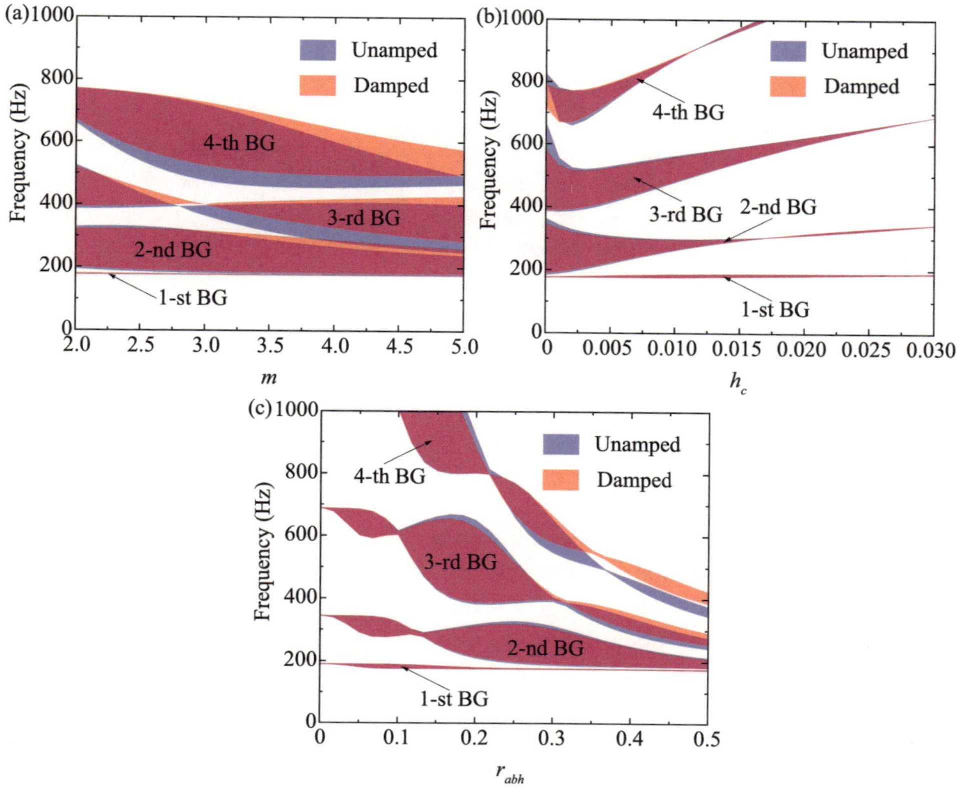

Figure 5.
Band gaps changing with (a) the ABH order, (b) the central thickness, and (c) the ABH radius.

former case. Due to the added stiffness of the layer, the BGs however occur at higher frequencies.

Different from the ABH order, the BGs are very sensitive to the central thickness. As graphed in **Figure 5b**, generally, the BGs become very narrow as the central thickness increases from 0^+ to $h_{uni} = 0.03$ m. It seems that the central thickness of the ABH needs to be small enough for wide BGs, whose local resonance is more significant. Howbeit, it is not realistic because truncation always exists. Moreover, we observe that for the 4-th BG there is an inflection point near $h_c = 0.002$ m, indicating that the optimal thickness can be found in this range. For comparison, the effects of the damping layer are also characterized in **Figure 5b**, but it seems not important to the BGs. However, for very small r_{abh} the added mass of the damping layer is more dominant than its added stiffness. This is why the BGs for the damped shell are lower than those for the undamped shell.

Finally, the effects of the ABH radius are computed and illustrated in **Figure 5c**. We consider the ABH radius varying from 0 to $L_{cell}/2 = 0.5$ m. From the figure, we can see that the BGs move to low-frequency range very fast as r_{abh} grows. Particularly, the BGs are prone to inverse when the radius is not very large ($r_{abh} < 0.4$ m). The 2-nd BG inverses at $r_{abh} = 0.14$ m, while the 3-rd BG has double inversions respectively at $r_{abh} = 0.1$ m and $r_{abh} = 0.32$ m, whereas the 4-th BG contains at least two inversions at $r_{abh} = 0.22$ m and $r_{abh} = 0.37$ m. When $r_{abh} > 0.4$ m, the BGs become stable and progressively move to lower frequencies. Again, the influence of the damping layer is also characterized in **Figure 5c**. On the one hand, it is inspected that the damping layer has significance merely for a large ABH radius. On the other, the BGs locate at higher frequencies because of added stiffness, as reported in **Figure 5a**.

3. Sound radiation from finite periodic ABH shells

3.1 Radiation theory for cylindrical shells

Once the radial displacement $w(x',\theta')$ (here $\theta'=y'/R$) is obtained, the normal velocity to the surface can be further expressed as $v_w(x',\theta')=\mathrm{j}\omega w(x',\theta')$. For a baffled cylindrical shell, the radiated sound pressure of an arbitrary external point (r,θ,x) can be analytically obtained by [29, 30].

$$p(x,\theta,r)=\frac{\mathrm{j}\rho_a\omega}{4\pi^2}\int_{-\pi}^{\pi}\int_{-a}^{a}v_w(x',\theta')\sum_{n=-\infty}^{+\infty}\cos\left[n(\theta-\theta')\right]\times$$
$$\int_{-\infty}^{+\infty}\frac{\exp\left[\mathrm{j}k_x(x-x')\right]}{k_yR}\frac{\mathrm{H}_n^{(1)}(k_yr)}{\mathrm{H}_n^{(1)'}(k_yR)}\mathrm{d}k_x\mathrm{d}x'R\mathrm{d}\theta', \tag{17}$$

where ρ_a represents the density of air, k_x and $k_y=\left(k^2-k_x^2\right)^{0.5}$ symbolize the wavenumber components in the x and the y direction, respectively, and k stands for the total sound wavenumber. Here $\theta=y/R$ stands for the circumferential angle. $\mathrm{H}_n^{(1)}$ indicates the n-th Hankel function of the first kind, and $\mathrm{H}_n^{(1)\prime}$ denotes its first derivative with respect to the argument k_yR.

For numerical estimation, the cylindrical shell can be segmented into N elementary radiators, with each surface area ΔS. The surface velocity can be assembled as a vector $\mathbf{v}_w$, which can be further used to calculate the sound pressure vector on the cylindrical surface

$$\boldsymbol{p}_{N\times 1}=\boldsymbol{Z}_{N\times N}\boldsymbol{v}_{N\times 1}, \tag{18}$$

where $\boldsymbol{Z}$ represents the acoustic impedance matrix, with entries

$$Z_{ij}=\frac{\mathrm{j}\rho_a\omega\Delta S}{2\pi^2}\sum_{n=0}^{+\infty}\varepsilon_n\cos\left[n\left(\theta_i-\theta_j\right)\right]\int_0^{+\infty}\frac{\cos\left[k_x\left(x_i-x_j\right)\right]}{k_ya}\frac{\mathrm{H}_n^{(1)}(k_yr)}{\mathrm{H}_n^{(1)'}(k_yR)}\mathrm{d}k_x, \tag{19}$$

here ε_n is a normalized coefficient, and it is given by

$$\varepsilon_n=\begin{cases}1, & n=0,\\ 2, & n>0.\end{cases} \tag{20}$$

Next, we can write the sound power as

$$W_s=\boldsymbol{v}^{\mathrm{H}}\boldsymbol{R}\boldsymbol{v}, \tag{21}$$

where the superscript H stands for the Hermite transpose and $\boldsymbol{R}=\frac{\Delta S}{2}\,\mathrm{Re}\left(\boldsymbol{Z}|_{r=R}\right)$ is the radiation resistance matrix (real symmetric and positive-definite). The sound radiation efficiency can be further obtained by

$$\sigma=\frac{W_s}{\rho_a c_a N\Delta S\left\langle v_w^2\right\rangle_{overall}}, \tag{22}$$

where $\left\langle v_w^2\right\rangle_{overall}$ represents the mean square velocity (MSV) over the whole surface of the ABH cylindrical shell.

3.2 Sound radiation from unstiffened ABH shells

Now we investigate the sound radiation from a finite unstiffened shell containing five ABH cells. Based on the reconstructed GEM presented in Section 2.1, we cancel the axial periodic conditions, namely Eqs. (12) and (13), such that the vibration field of the finite ABH shell can be characterized. It is well-known that there is a cut-on frequency for an ABH, which is mainly determined by its size. Only beyond this frequency (wavelength smaller than the ABH size, $2r_{abh}$), the wave can be trapped and the ABH effect can be triggered. According to **Table 1**, the radius cut-on frequency $f_r = \frac{\pi h_{uni}}{4r_{abh}^2}\sqrt{\frac{E}{3\rho(1-\nu^2)}} = 282$ Hz.

As plotted in **Figure 6a**, the radial mean square velocity (MSV) on the surface of the ABH shell is compared to that of the uniform (UNI) shell having the same damping layer configuration. From the figure, it can be seen that in the BGs, the vibration of the ABH shell is very low. In the passbands, the vibration beyond $f_r = 282$ Hz is stronger but never exceeds the MSV of the reference shell. Particularly, the average reduction reaches $\Delta MSV = 10$ dB. Looking at **Figure 6b**, the sound power level (SWL) is also very small in the BGs. Even for the passbands, the SWL is effectively suppressed, with an average reduction approaching $\Delta SWL = 15$ dB. Then, we examine the radiation efficiency of the ABH shell. As shown in **Figure 6c**, the radiation efficiency of the uniform shell grows to the maxima in the vicinity of the critical frequency $f_c = 390$ Hz. After embedding the ABHs, however, the radiation efficiency is significantly impaired, almost in the whole frequency band of

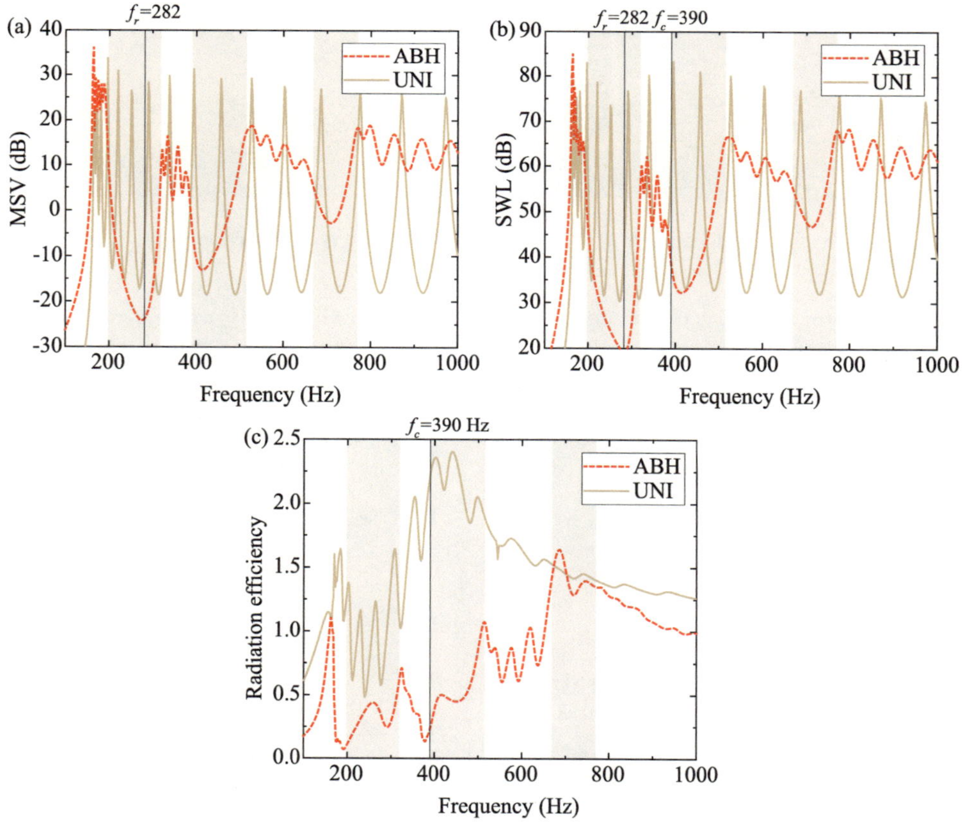

Figure 6.
Comparison of the (a) mean square velocity (MSV), (b) sound power level (SWL), and (c) radiation efficiency, between the ABH and uniform shells. The shaded areas stand for the BGs.

interest. Careful readers may notice that the radiation efficiency is very small in the 2-nd BG (the 1-st BG is not shown here because it is too narrow), but in the latter BGs, it becomes larger. This is due to the weakening of the local resonance effect at higher frequencies where the uniform portions start to activate (see **Figure 4**).

To clearly manifest the characteristics of the ABH shell, we have further calculated the vibration field at each frequency. As illustrated in **Figure 7**, compared to the uniform shell whose distribution of vibration nodes is very regular to location and frequency (over the ring frequency 173 Hz), the vibration in the BGs is obviously isolated as propagating in the axial direction. For frequencies outside of the BGs, the amplitude of the displacement is also clearly reduced to the right direction. Specifically, we choose 287 Hz and 340 Hz as two representative frequencies in the BG and the passband, respectively, for demonstrating the effectiveness of the ABH shell. In **Figure 8a**, it is clearly seen that the local resonance effect in the ABHs is very strong, such that vibration can be substantially stopped when compared to the uniform shell. While in **Figure 8b** we can see that the wave can be transmitted to the whole shell, but the amplitude is very small because of the highly efficient damping effect by the ABH + damping layer configuration.

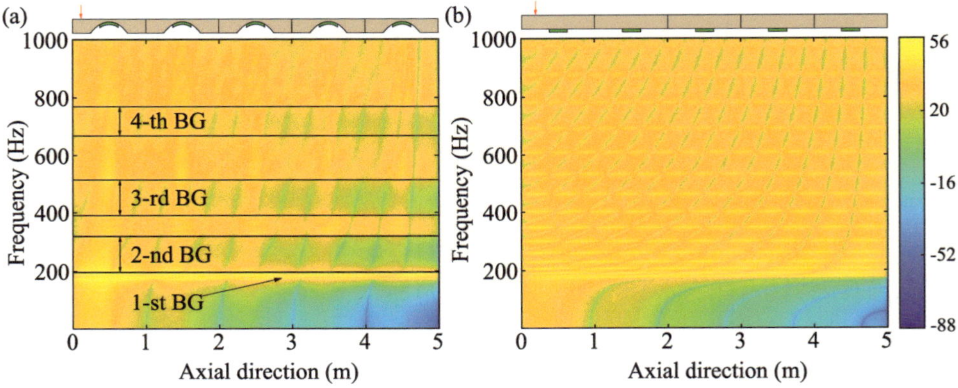

Figure 7.
Displacements (w_{ref} = 1 m) of (a) the ABH shell and (b) the uniform shell, changing with frequency.

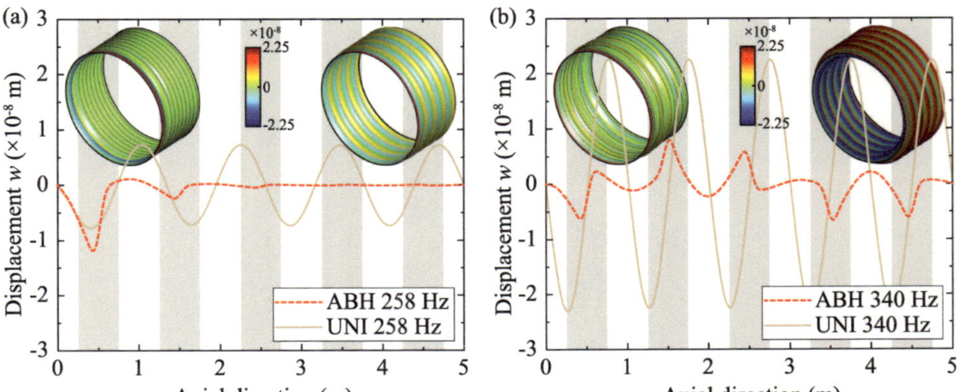

Figure 8.
Forced vibration shapes for the finite shell having five cells at (a) 258 Hz (at the center of the second bandgap), (b) 340 Hz (in the passband), compared to a uniform cylindrical shell having the same damping layer configuration. The shaded areas represent the ABH portions.

3.3 Sound radiation from stiffened ABH shells

Note that the embedded ABHs may obviously reduce the structural stiffness of the shell, therefore, the stiffeners (see **Figure 2d**) can be utilized to alleviate this issue. In this section, we first investigate the effects of the stiffener number, then study that of the stiffener width.

For simplicity, let us first consider three cases. That is, the stiffener number $N = \{8, 16, 32\}$, while the width of each stiffener keeps $W = 4h_{uni}$. As shown in **Figure 9a**, when adding eight stiffeners the vibration level is very similar to the ABH shell (without stiffener), except for the frequencies in the BGs. Increasing the stiffener number will further deteriorate the vibration in the BGs, but it seems that the MSV level is similar to that of the unstiffened one. This means that the combination of the ABH and stiffeners not only results in a more rigid structure (compared to the pure ABH one) but also maintains the overall damping effect. Similar results can be found for the radiated SWL (see **Figure 9b**). The existence of the stiffeners almost merely increases the SWL in the BGs, yet in the passbands, the stiffened ABH shell has a similar SWL to that of the unstiffened one. Furthermore, the radiation efficiency is checked (see **Figure 9c**). From the figure, we can see that after inserting stiffeners the radiation efficiency in the 2-nd (the 1-st BG is not shown here) is very small and, that for the case of 32 stiffeners the efficiency is the lowest over 700 Hz. In general, the stiffeners will not obviously degrade the reduction ability of the ABH shell.

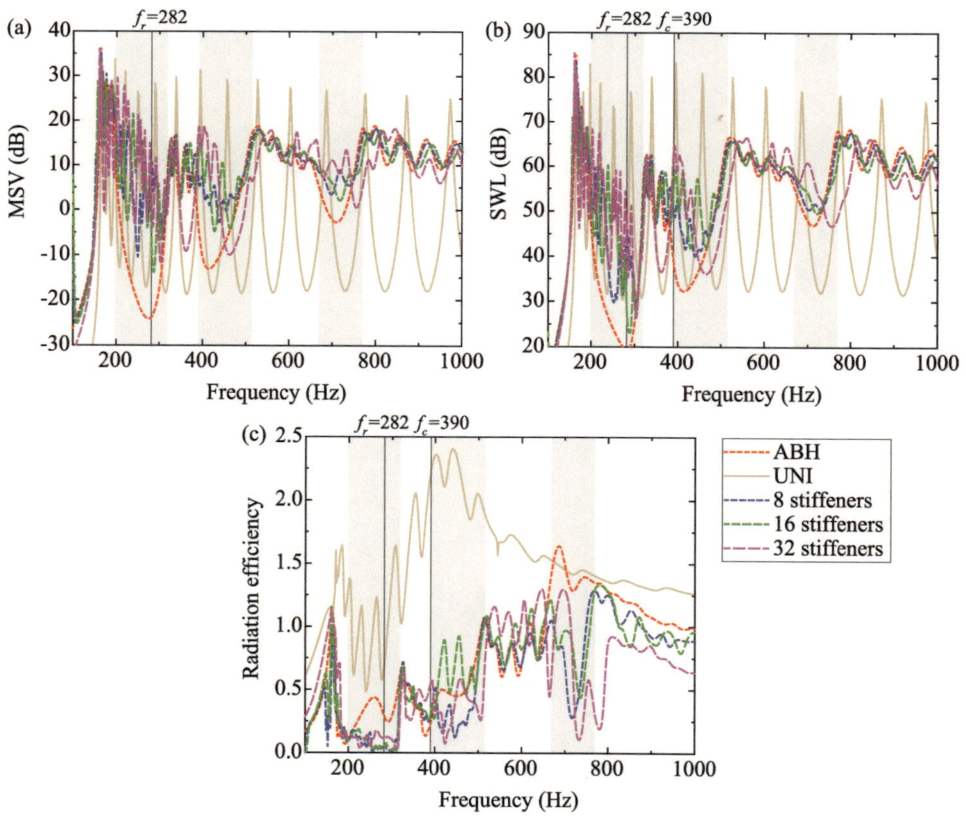

Figure 9.
(a) Mean square velocity (MSV), (b) sound power level (SWL), and (c) radiation efficiency, for the cylindrical shell having a different number of stiffeners with the same width $W = 4h_{uni}$.

Now we keep the 16 stiffeners in the circumferential direction for each cell but modify the width from $W = 4h_{uni}$ to $W = 8h_{uni}$, then to $W = 16h_{uni}$. Similar to the above results, from **Figure 10** it can be observed that increasing the stiffener width mainly intensifies the MSV and SWL in the BGs, while their values in the passbands almost do not change. Particularly, the radiation efficiency of the stiffened shells is obviously lower than the ABH shell at frequencies greater than 650 Hz. Even large stiffener width will enlarge the vibration and the radiated sound power at low frequencies (e.g., at 180 Hz), the high-frequency performance is still plausible and the structure is more rigid.

For the purpose of illustration, we have figured out the normal velocity and sound pressure distributions on the finite shells at 340 Hz (in passband). The former is graphed in **Figure 11** while the latter in **Figure 12**. From **Figure 11**, we can see that the vibration level of the uniform shell is very strong (**Figure 11a**), but the ABHs can help reduce the overall vibration, with only strong vibrations in the ABH portions where the damping layer is very effective (**Figure 11b**). After adding stiffeners (**Figure 11c1–d3**), the vibration in the ABH areas is intensified compared to **Figure 11b**. The sound pressure distributions in **Figure 12** display that the ABH and stiffened ABH shells can also effectively reduce the sound pressure, compared to the reference uniform shell. The existence of the stiffener makes the distributions no longer axially symmetrical, except for **Figure 12c3** where a lot of small and distributed stiffeners are imposed.

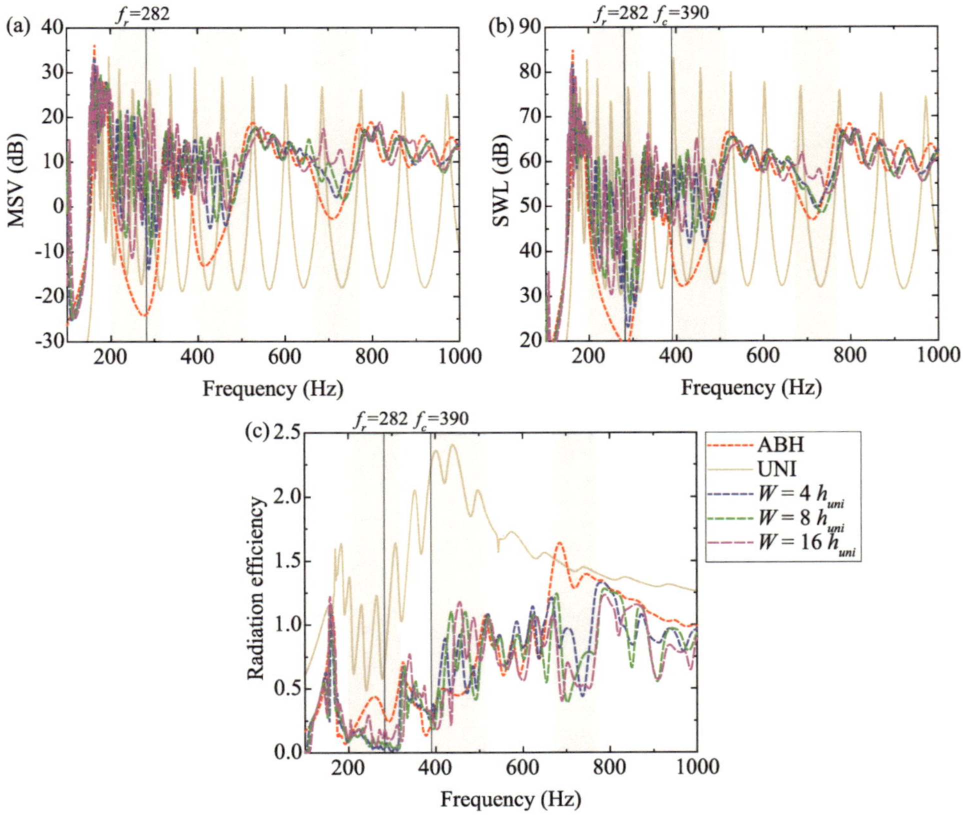

Figure 10.
(a) Mean square velocity (MSV), (b) sound power level (SWL), and (c) radiation efficiency, for the cylindrical shell having 16 stiffeners with different widths.

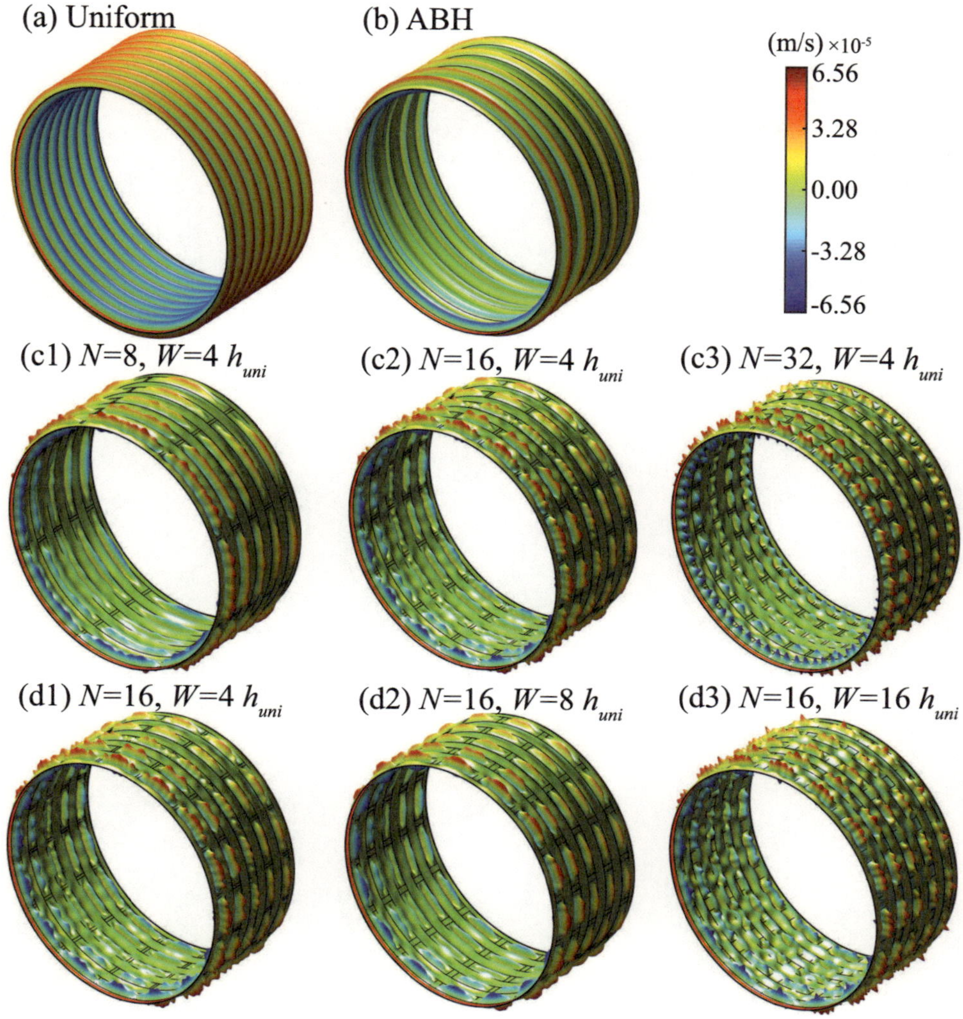

Figure 11.
Surface normal velocity distributions at 340 Hz. (a) Uniform shell, (b) ABH shell without stiffener, (c1)-(c3) stiffened ABH shell with the different number of stiffeners of the same width, (d1)-(d3) stiffened ABH shell with 16 stiffeners having different widths. The red circles stand for the ring force.

4. Conclusions

In the current chapter, the acoustic black hole (ABH) effect is concerned and applied to cylindrical structures. By reducing thickness following the power law, the wave velocity is substantially slowed and the wavenumber is increased when it propagates to the ABH center, where the damping layer is very efficient to consume vibrational energy. The focus is placed on reducing the sound emission from cylindrical shells, which can be found in many fields, via embedding periodic ABHs. First, the reconstructed Gaussian expansion method (GEM) is presented to characterize infinite periodic ABH shells. The band gaps (BGs), induced by the locally resonant effect in the ABH area, are investigated, together with the influence of the ABH parameters. Next, the sound radiation model for finite periodic cylindrical shells is developed. Numerical results show that the periodic ABHs can both reduce the vibration and sound power, relying on two mechanisms—(i) the BGs for

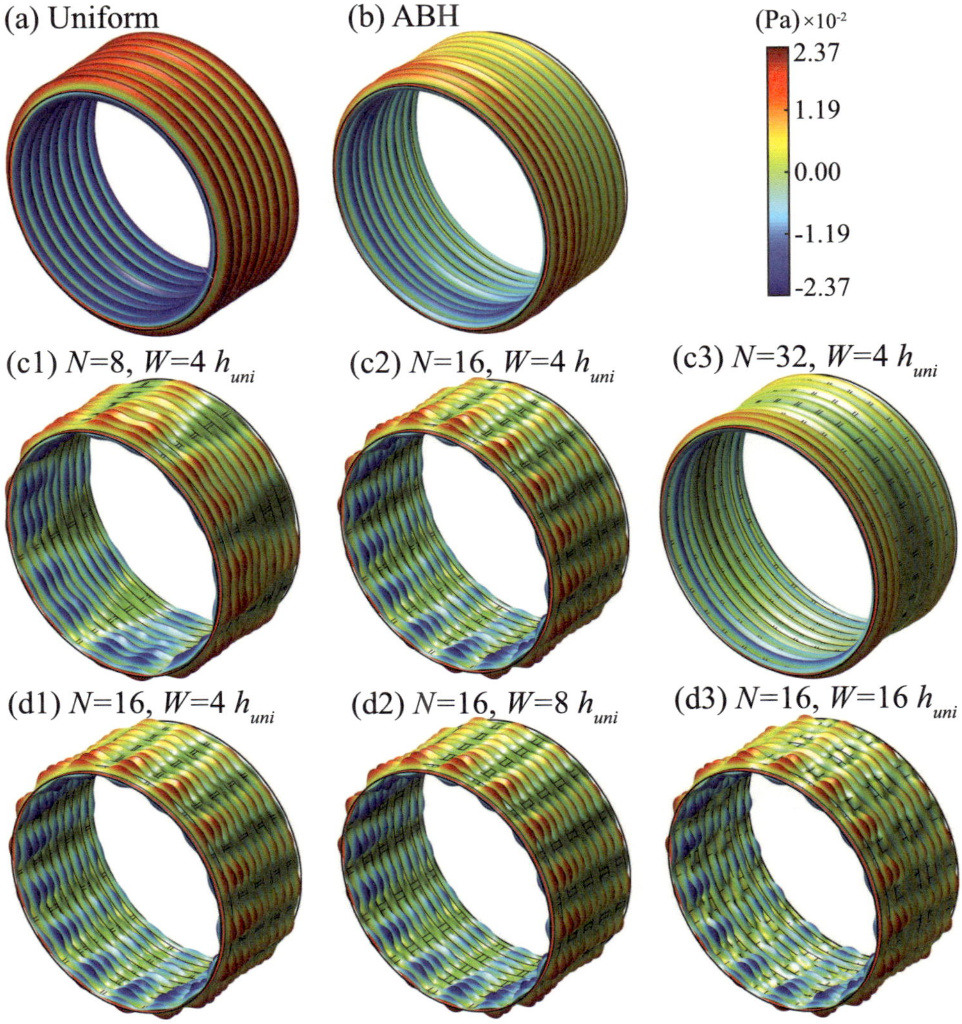

Figure 12.
Surface sound pressure distributions at 340 Hz. (a) Uniform shell, (b) ABH shell without stiffener, (c1)-(c3) stiffened ABH shell with the different number of stiffeners of the same width, (d1)-(d3) stiffened ABH shell with 16 stiffeners having different widths. The red circles stand for the ring force.

isolating vibrations and (ii) the damping effects for energy consumption. The inclusion of stiffeners not only strengthens the structural stiffness but also keeps the vibration and sound power level as the pure ABH one in the passbands.

Acknowledgements

This work was supported by the National Natural Science Foundation of China (grant number 11704314) and the China Postdoctoral Science Foundation (grant numbers 2018M631194 and 2020T130533).

Conflict of interest

The authors declare no conflict of interest.

Author details

Jie Deng* and Nansha Gao
Key Laboratory of Ocean Acoustic and Sensing, School of Marine Science and Technology, Northwestern Polytechnical University, Xi'an, China

*Address all correspondence to: dengjie@nwpu.edu.cn

References

[1] Krylov VV, Tilman FJBS. Acoustic black holes for flexural waves as effective vibration dampers. Journal of Sound and Vibration. 2004;**274**(3): 605-619. DOI: 10.1016/j.jsv.2003.05.010

[2] Krylov VV. A new type of vibration damper based on flexural wave propagation in laminated wedges of powerlaw profile. The Journal of the Acoustical Society of America. 2001; **110**(5):2654-2654. DOI: 10.1121/1.4777028

[3] Krylov VV, Winward RETB. Experimental investigation of the acoustic black hole effect for flexural waves in tapered plates. Journal of Sound and Vibration. 2007;**300**(1): 43-49. DOI: 10.1016/j.jsv.2006.07.035

[4] Deng J, Zheng L, Zeng P, Zuo Y, Guasch O. Passive constrained viscoelastic layers to improve the efficiency of truncated acoustic black holes in beams. Mechanical Systems and Signal Processing. 2019;**118**:461-476. DOI: 10.1016/j.ymssp.2018.08.053

[5] Zeng P, Zheng L, Deng J, Elsabbagh A, Xiang S, Yan T, et al. Flexural wave concentration in tapered cylindrical beams and wedge-like rectangular beams with power-law thickness. Journal of Sound and Vibration. 2019;**452**:82-96. DOI: 10.1016/j.jsv.2019.04.002

[6] Li H, Touz C, Pelat A, Gautier F, Kong X. A vibro-impact acoustic black hole for passive damping of flexural beam vibrations. Journal of Sound and Vibration. 2019;**450**:28-46. DOI: 10.1016/j.jsv.2019.03.004

[7] Deng J, Guasch O, Zheng L. Ring-shaped acoustic black holes for broadband vibration isolation in plates. Journal of Sound and Vibration. 2019; **458**:109-122. DOI: 10.1016/j.jsv.2019.06.017

[8] Ma L, Zhang S, Cheng L. A 2D Daubechies wavelet model on the vibration of rectangular plates containing strip indentations with a parabolic thickness profile. Journal of Sound and Vibration. 2018; **429**:130-146. DOI: 10.1016/j.jsv.2018.04.042

[9] Tang L, Cheng L, Ji H, Qiu J. Characterization of acoustic black hole effect using a one-dimensional fully-coupled and wavelet-decomposed semi-analytical model. Journal of Sound and Vibration. 2016;**374**:172-184. DOI: 10.1016/j.jsv.2016.03.031

[10] Jeon W, Lee JY. Vibration damping using a spiral acoustic black hole. The Journal of the Acoustical Society of America. 2017;**141**(5):3644-3644. DOI: 10.1121/1.4987872

[11] Tang L, Cheng L. Enhanced acoustic black hole effect in beams with a modified thickness profile and extended platform. Journal of Sound and Vibration. 2017;**391**:116-126. DOI: 10.1016/j.jsv.2016.11.010

[12] Fu Q, Du X, Wu J, Zhang J. Dynamic property investigation of segmented acoustic black hole beam with different power-law thicknesses. Smart Materials and Structures. 2021; **30**(5):055001. DOI: 10.1088/1361-665x/abed32

[13] Bowyer EP, Krylov VV. Experimental study of sound radiation by plates containing circular indentations of power-law profile. Applied Acoustics. 2015;**88**:30-37. DOI: 10.1016/j.apacoust.2014.07.014

[14] Li X, Ding Q. Sound radiation of a beam with a wedge-shaped edge embedding acoustic black hole feature. Journal of Sound and Vibration. 2019; **439**:287-299. DOI: 10.1016/j.jsv.2018.10.009

[15] Ma L, Cheng L. Sound radiation and transonic boundaries of a plate with an acoustic black hole. The Journal of the Acoustical Society of America. 2019; **145**(1):164-172. DOI: 10.1121/1.5081680

[16] Deng J, Guasch O, Maxit L, Zheng L. Transmission loss of plates with multiple embedded acoustic black holes using statistical modal energy distribution analysis. Mechanical Systems and Signal Processing. 2021;**150**:107262. DOI: 10.1016/j.ymssp.2020.107262

[17] Deng J, Zheng L. Noise reduction via three types of acoustic black holes. Mechanical Systems and Signal Processing. 2022;**165**:108323. DOI: 10.1016/j.ymssp.2021.108323

[18] Deng J, Guasch O, Zheng L. Reconstructed Gaussian basis to characterize flexural wave collimation in plates with periodic arrays of annular acoustic black holes. International Journal of Mechanical Sciences. 2021; **194**:106179. DOI: 10.1016/j. ijmecsci.2020.106179

[19] Deng J, Zheng L, Gao N. Broad band gaps for flexural wave manipulation in plates with embedded periodic strip acoustic black holes. International Journal of Solids and Structures. 2021; **224**:111043. DOI: 10.1016/j. ijsolstr.2021.111043

[20] Deng J, Zheng L, Guasch O. Elliptical acoustic black holes for flexural wave lensing in plates. Applied Acoustics. 2021;**174**:107744. DOI: 10.1016/j.apacoust.2020.107744

[21] Deng J, Guasch O, Zheng L, Song T, Cao Y. Semi-analytical model of an acoustic black hole piezoelectric bimorph cantilever for energy harvesting. Journal of Sound and Vibration. 2021;**494**:115790. DOI: 10.1016/j.jsv.2020.115790

[22] Deng J, Guasch O, Zheng L. A semi-analytical method for characterizing vibrations in circular beams with embedded acoustic black holes. Journal of Sound and Vibration. 2020;**476**: 115307. DOI: 10.1016/j.jsv.2020.115307

[23] Deng J, Guasch O, Maxit L, Zheng L. Vibration of cylindrical shells with embedded annular acoustic black holes using the Rayleigh-Ritz method with Gaussian basis functions. Mechanical Systems and Signal Processing. 2021; **150**:107225. DOI: 10.1016/j. ymssp.2020.107225

[24] Deng J, Guasch O, Maxit L, Zheng L. Reduction of Bloch-Floquet bending waves via annular acoustic black holes in periodically supported cylindrical shell structures. Applied Acoustics. 2020;**169**:107424. DOI: 10.1016/j.apacoust.2020.107424

[25] Deng J, Guasch O, Maxit L, Zheng L. Annular acoustic black holes to reduce sound radiation from cylindrical shells. Mechanical Systems and Signal Processing. 2021;**158**:107722. DOI: 10.1016/j.ymssp.2021.107722

[26] Deng J, Zheng L, Guasch O, Wu H, Zeng P, Zuo Y. Gaussian expansion for the vibration analysis of plates with multiple acoustic black holes indentations. Mechanical Systems and Signal Processing. 2019;**131**:317-334. DOI: 10.1016/j.ymssp.2019.05.024

[27] Sivasubramonian B, Rao GV, Krishnan A. Free vibration of longitudinally stiffened curved panels with cutout. Journal of Sound and Vibration. 1999;**226**(1):41-55. DOI: 10.1006/jsvi.1999.2281

[28] Tang L, Cheng L. Broadband locally resonant band gaps in periodic beam structures with embedded acoustic black holes. Journal of Applied Physics. 2017;**121**(19):194901. DOI: 10.1063/ 1.4983459

[29] Stepanishen PR. Radiated power and radiation loading of cylindrical

surfaces with nonuniform velocity distributions. The Journal of the Acoustical Society of America. 1978; **63**(2):328-338. DOI: 10.1121/1.381743

[30] Sun Y, Yang T, Chen Y. Sound radiation modes of cylindrical surfaces and their application to vibro-acoustics analysis of cylindrical shells. Journal of Sound and Vibration. 2018;**424**:64-77. DOI: 10.1016/j.jsv.2018.03.004

Chapter 6

Trailing Edge Bluntness Noise Characterization for Horizontal Axis Wind Turbines [HAWT] Blades

Satya Prasad Maddula, Vasishta Bhargava Nukala, Swamy Naidu Neigapula Venkata, Chinmaya Prasad Padhy and Rahul Samala

Abstract

Wind turbine noise is becoming a critical issue for many offshore and land-based wind projects. In this work, we analyzed trailing edge bluntness vortex shedding noise source for a land-based turbine of size 2 MW and blade span of 38 m using original Brooks Pope and Marcolini (BPM)and modified BPM noise model. A regression-based curve fitting approach has been implemented to predict the shape function in terms of thickness to chord ratio of aerofoils used for blade. For trailing edge height of 0.1% chord, computations for sound power level were done at wind speed of 8 m/s, 17 RPM. The results showed that present approach for thickness correction predicts the noise peak of ~78dBA at f ~ 10 kHz which is ~15dBA lower than that predicted from original BPM. The results were also validated using experiment data from GE 1.5sle, Siemens 2.3 MW turbines with blade lengths between 78 m and 101 m which agreed within 2% at high frequencies, f > 5 kHz. In addition, results from present approach for trailing edge bluntness noise agreed well with modified BPM by Wei et al. at high frequencies, f ~ 10 kHz where it becomes dominant. The slope of noise curves from present approach, and modified BPM methods are lower when compared with original BPM.

Keywords: Noise, blades, trailing edge, turbulent boundary layer, sound power

1. Introduction

Wind power is growing at exponential rate with installed wind power capacity reached more than 500 GW globally. By far the cheapest source of energy generation among all renewable energy technologies is wind power. As more wind power projects are installed, a growing concern of noise emissions from wind turbine blades is increasing due to adverse health effects on inhabitants living near wind farms [1]. Many policy makers are considering this issue seriously as noise generated from wind turbines is an impediment to the growth of wind energy growth. Modern megawatt scale turbines have large rotor diameter of size 100 m and above which contribute to the overall noise levels and cause annoyance for people living near wind farms. **Figure 1** depicts the evolution of size of horizontal axis wind turbines over the period of forty years. Size of rotor diameter ranges from 17 m to ~165 m with its nominal power range between 70 kW to 6 MW.

Airfoil self-noise from wind turbines with longer blades have higher tip speeds and produce high aerodynamic noise. Studies by several researchers have found that most of the broadband aerodynamic noise emissions occur due to trailing edge source from rotating blades such as from helicopter, wind turbines and compressors [2–5]. However when the blades become thicker, the trailing edge bluntness source also dominates between moderate to high frequency range in noise spectra. Airfoil self-noise prediction models developed by Brooks Pope and Marcolini (BPM) have been studied and improved by several researchers [6–8]. One of the recent improvements in the trailing edge bluntness noise predictions was done by Wei et al. (2016) who applied numerical techniques and correlated their results with field experiments measured for Siemens 2.3 MW wind turbine blade. They also used computational aero-acoustic (CAA) method to compute the trailing edge bluntness noise level from NACA 0012 airfoil with finite thickness for consistent validation of results obtained from BPM semi empirical noise prediction model and measured noise data. In addition, NACA 63–418 with two different variants of trailing edge shapes were studied to compare the noise spectra. They modified the generalized shape function proposed by original BPM model and made it independent of the solid angle formed between the trailing edge surfaces of airfoil to investigate the effect of the shape function on the trailing edge tonal noise peak produced in the high frequency region of sound spectra. In the present study, we investigate the shape function used by BPM model for predicting the trailing edge bluntness noise source but also apply regression approach to improve the bluntness peak at the high frequency region of noise spectra. To the best of authors knowledge, regression approach has not been implemented before to study the effects of trailing edge tonal noise source for wind turbine blades. In Section 2, we describe the trailing edge bluntness noise method developed by BPM along with present formulation. In Section 3, geometry model of wind turbine blade used in study is described along with IEC 61400–11 standards for measurements of acoustic emissions for wind turbines. Computational assumptions are described for the generating aerodynamic flow field by means of BEM which is coupled to the noise solver for predicting sound power level. The noise solver for the trailing edge bluntness source is

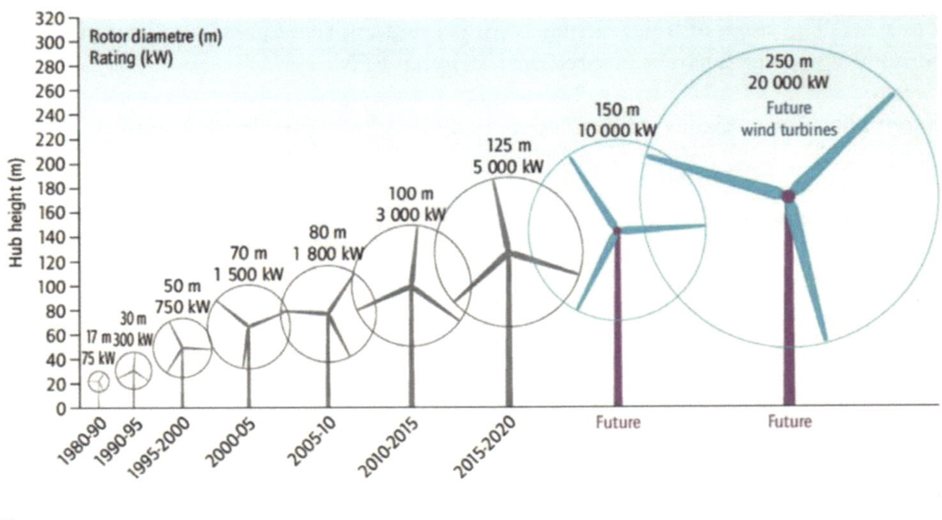

Figure 1.
Illustration of size of horizontal axis wind turbines (HAWT) and its evolution over a period of forty years.

developed based on the original BPM model along with its improvements proposed by [9]. Regression method is then applied on trailing edge noise shape function based on the coefficients obtained for the modified trailing edge height of the airfoils along the span wise direction of blade. In Section 4, present results for trailing edge bluntness noise source are compared to those obtained from original BPM, modified BPM by [9]. Overall 1/3rd octave band sound power level for the 2 MW wind turbine with a blade length of 38 m are computed and validated with experiment data of the GE 1.5sle, Siemens SWT 2.3 MW with 93 m, 95 m, and 101 m versions of turbines. Finally, conclusions are presented based on the results obtained for the original BPM, modified BPM and present correction function for airfoil thickness to chord ratio.

2. Methods

2.1 BPM model: trailing edge bluntness vortex shedding

Flow around wind turbine blades can be considered often as incompressible and low Mach number for most utility scale wind turbines. Even though they operate in environments where the effects of air density and wind shear on power production are significant, aerodynamic noise generation from wind turbine blades becomes important when the blade tip speed range between 0.1 and 0.3 Mach number. As length of blade is increased, the sound radiation from blades depends not only upon aerofoil geometry, local angle of attack for the aerofoils but also the rotational speed of rotor. One of the noise mechanisms from blades occurs due to periodic vortex shedding from suction side of trailing edge surface when the turbulent boundary layer flow interacts with blade surface and contributes to a monotonic peak in high frequency region of noise spectrum. For a given flow condition i.e., Reynolds number and Mach number along the span wise direction of the blade strongly affects the overall noise levels as well as the tonal noise production. Typically, the noise amplitudes increase with increase in flow Mach number and Reynolds number of order 8×10^6.

Vortex shedding is aerodynamic phenomenon observed on both streamlined and bluff bodies such as an aerofoil or a cylinder and becomes dominant when there exists an adverse pressure gradient within the boundary layer which causes a relative difference in the flow velocities between the surface and free stream flow conditions. According to BPM model, the trailing edge vortex shedding occurs when the turbulent boundary layer displacement thickness is at least 30% higher than characteristic dimension of source [6, 10]. In addition, flow conditions such as angle of attack, Reynolds number and Mach number affect the aerodynamic lift and drag force characteristics of an aerofoil. It can be noted that for low angle of attack and attached flows, vortex shedding from trailing edge occurs rapidly and produces unsteady lift which often result in higher noise generation [11]. The lift and drag coefficient at high angle of attack also increases rapidly but reach maximum values near stall angle of attack. For aerofoils with finite trailing edge thickness, and at stall angle of attack, the vortex shedding phenomenon is reduced due to turbulent boundary layer separation near the trailing edge. Beyond the stall angle of attack, significant reduction of lift can be observed and hence vortex noise from aerofoils is also reduced. BPM model predicts noise radiation from aerofoils using relative velocity and angle of attack as primary inputs and computes the turbulent boundary layer data for suction and pressure sides of aerofoil. This data varies according to the thickness of trailing edge of aerofoil as well as the chord length of aerofoil. As the thickness to chord increases, the turbulent boundary layer on the suction side of aerofoil becomes less stable and tend to shed vortices rapidly. For a rotating wind turbine blade, the vortex shedding occurrence happens at faster rate which leads to the massive flow separation near the tip of blade due to centrifugal force action on the flow. The separated flow appears as wake which has lower velocity compared to free stream flow condition and contributes to aerodynamic noise. Further, according to this method, the strength of this source is approximated using the spectral functions, G_4 and G_5 which are functions of ratio of trailing edge thickness and average turbulent boundary layer displacement thickness from pressure and suction sides of aerofoil as given by Eq. (5). Hence, it is needed to compute the spectral functions G_4 and G_5 as given by Eq. (6)–(8). G_4 represents the narrowband peak in spectra and G_5 is used to determine the broadband overall shape of spectra which is dependent on Strouhal number, St''' and St_{peak}'''.

The two spectral functions $(G_5)_{\varphi=14}$ and $(G_5)_{\varphi=0}$ are solid angles which are determined using the symmetric NACA 0012 aerofoil experiments and given by Eq. from (76) to (82) in (Brooks et al., 1989). As mentioned by [6, 10], the bluntness vortex shedding source appears as tonal peak in the overall noise spectra and becomes dominant near 10 kHz masking other self-noise mechanisms. It must be noted that the functional parameters in Eq. (1) are expressed in terms of the flow angle of attack, bluntness ratio h/δ^*, for aerofoil at moderate to high Reynolds number; at the same time, they show the dependence of Mach number, $M^{5.5}$. The noise levels are also found to vary with the span segment length of aerofoil, L and inverse square of the distance between source and receiver, r^2_e as given in Eq. (1). The Strouhal number for this type of source is defined according to Eq. (2) where h is the height of trailing edge. It must be noted that at moderate Reynolds number and for subsonic Mach number flows, the chord Reynolds number and turbulent boundary layer thickness and displacement thicknesses for zero and non-zero angle of attack are evaluated using Eq. (5) and Eq. (16) given in [6]. The 1/3rd octave sound pressure for this source is approximated using the Eq. (1). The narrowband tonal peak is given by function G_4 and expressed using Eqs. (6) and (7).

Function G_5 is calculated using ratio of trailing edge thickness to average boundary layer displacement thickness and sloping angle, φ between 0° to 14° given by Eq. (78) and Eq. (79) found in [6] where φ is the angle between the sloping surfaces near trailing edge of aerofoil and δ^*_p and δ^*_s are the pressure and suction side turbulent boundary layer displacement thickness, and h is the trailing edge height. The empirical equations used to determine the pressure and suction side displacement thicknesses for zero and non-zero angle of attack for symmetric aerofoils are given in [6]. They are found to be dependent upon the local angle of attack and chord Reynolds number. For an aerofoil, it is expressed in terms of the turbulent boundary layer displacement thicknesses for the pressure and suction side. This source also uses the high frequency directivity function like turbulent boundary layer trailing edge noise and given by the Eq. (9).

$$SPL_{\text{Blunt}} = 10\log\frac{hM^{5.5}LD_{\text{h}}}{r_{\text{e}}^2} + G_4\left(\frac{h}{\delta^*_{\text{avg}}},\varphi\right) + G_5\left(\frac{h}{\delta^*_{\text{avg}}},\varphi,\frac{St'''}{St'''_{\text{peak}}}\right), \tag{1}$$

$$St''' = \frac{fh}{U}, \tag{2}$$

$$St'''_{\text{peak}} = \frac{0.212 - 0.0045\varphi}{1 + 0.235\left(\frac{h}{\delta^*_{\text{avg}}}\right)^{-1} - 0.0132\left(\frac{h}{\delta^*_{\text{avg}}}\right)^{-2}}, \quad \text{for } \frac{h}{\delta^*_{\text{avg}}} \geq 0.2, \tag{3}$$

$$St'''_{\text{peak}} = 0.1\left(\frac{h}{\delta^*_{\text{avg}}}\right) + 0.095 - 0.00243\varphi, \quad \text{for } \frac{h}{\delta^*_{\text{avg}}} < 0.2, \tag{4}$$

$$\delta^*_{\text{avg}} = \frac{\delta^*_{\text{p}} + \delta^*_{\text{s}}}{2}, \tag{5}$$

$$G_4\left(\frac{h}{\delta^*_{\text{avg}}}, \varphi\right) = 17.5 \log \frac{h}{\delta^*_{\text{avg}}} + 157,5 - 1.114\varphi, \quad \textit{for} \ \frac{h}{\delta^*_{\text{avg}}} \leq 5, \tag{6}$$

$$G_4\left(\frac{h}{\delta^*_{\text{avg}}}, \varphi\right) = 169.7 - 1.114\varphi, \quad \textit{for} \ \frac{h}{\delta^*_{\text{avg}}} > 5, \tag{7}$$

$$G_5\left(\frac{h}{\delta^*_{\text{avg}}}, \varphi, \frac{St'''}{St'''_{\text{peak}}}\right) = (G_5)_{\varphi=0^\circ} + 0.0714\varphi\left[(G_5)_{\varphi=14^\circ} - (G_5)_{\varphi=0^\circ}\right], \tag{8}$$

$$D_H(\theta, \phi) = \frac{2\sin^2(1/2\theta)\sin^2\phi}{(1 + \text{M} \cdot \cos\theta)[1 + (\text{M} - \text{M}_{\text{C}})\cos\theta]^2}, \tag{9}$$

where θ, ϕ are the directivity angles between the source and receiver line aligned to blade span and chord direction with respect to the receiver position. M is the Mach number and M_c is the convective Mach number. h, is the trailing edge height. The denominator term in Eq. (9) represents the Doppler effect and convective amplification of acoustic waves produced at the trailing edge of aerofoil [6, 10, 12, 13]. It has been proven that for high values of Strouhal number or for the order greater than 2, the flow is dominated by turbulent boundary layer thickness and results in small scale flow instabilities [6, 14–16].

The Strouhal number and the shape functions vary with the shape of aerofoil, inflow velocity conditions and local angle of attack. Experiments conducted by [6] used a reference chord length for test aerofoil which was 30.86 cm and boundary tripping was done with help of 2 cm wide strip or grit applied at 15% chord length. Tripping of boundary layer resulted in reduction of the noise levels in certain frequency regions of sound spectrum [7, 8, 17]. For the present analysis, tripping of turbulent boundary layer has not been taken into consideration.

The maximum trailing edge height in BPM model aerofoil experiments was 2.5 mm which is ~0.8% of chord. For the present case of 38 m blade, it is 32.2 mm and corresponds to 1% chord, respectively.

2.2 Shape function and trailing edge thickness approximation proposed by Wei et al

As mentioned previously, trailing edge bluntness vortex shedding model was developed based on the experiment data obtained from NACA 0012. To account for the effects of vortex shedding noise levels, the geometry near the trailing edge requires an interpolation function essentially to approximate the height of trailing edge sloping surfaces. Standard solid angle was specified as $\psi = 14^\circ$ for a NACA 0012 aerofoil while for a flat plate it is $\psi = 0^\circ$. However, it must be noted that a wind turbine blade has finite thickness and varying camber along span direction. This led to erroneous predictions of the trailing edge noise levels. Hence [9] used a modified interpolation function for the trailing edge bluntness noise source and corrected the Eq. (1) using two additional functions viz. S_1 and S_2. S_1 is the shape function that is equivalent to the actual G_5 function and S_2 is the correction function for aerofoil thickness variation along the span wise direction of the blade given by Eqs. (10) and (11)

$$SPL_{\text{Blunt}} = 10\log_{10}\frac{hM^{5.7}LD_{\text{h}}}{r_{\text{e}}^{2}} + 20\left(1+M^{2}\right)\log_{10}\left(\frac{h}{\delta^{*}{}_{avg}}\right) + S_{1}\left(\frac{h}{\delta^{*}{}_{avg}},\frac{St}{St_{peak}}\right) + S_{2}\left(\frac{t}{c}\right) + K_{0} \tag{10}$$

$$S_{2} = 654.43\left(\frac{t}{c}\right)^{3} - 652.26\left(\frac{t}{c}\right)^{2} + 58.77\left(\frac{t}{c}\right) \tag{11}$$

Where, the constant, K_o is taken as 150 for $h/\delta^{*} < 0.2$ otherwise K_o is approximated as $150\text{–}20(h/\delta^{*} - 0.2)^{0.25}$. The Eq. (10) was modified in such a way that noise levels are not dependent of the solid angle formed at the trailing edge surfaces rather expressed as function of the bluntness height *h,* Mach number, *M,* and average of the boundary layer displacement thickness between suction and pressure sides of airfoil, δ^{*}_{avg}. Further, in the modified BPM for trailing edge bluntness, the sound pressure level is proportional $M^{5.7}$ instead of $M^{5.5}$. This change also demonstrates that the sound pressure for trailing edge bluntness source is sensitive to flow Mach number increments.

2.3 Modified thickness approximation using regression curve fitting

In the present study the basic shape function for the trailing edge angle is taken as from the original BPM model. However, for the shape function, G_5 the trailing edge angle is varied continuously between the blade root and tip section to account for differences in blade geometry. Since, the trailing edge sloping surfaces are proportional to the trailing edge height, a change in trailing edge angle parameter is retained in present noise computations while correction function for airfoil thickness, S_2 is modified in terms of thickness to chord ratio for each span segment of the blade similar to that proposed by [9]. One must note that coefficients in the modified function for thickness are obtained by regression and given by Eq. (12)

$$S_{2} = -0.02158\left(\frac{t}{c}\right)^{3} + 0.9518\left(\frac{t}{c}\right)^{2} - 13.38\left(\frac{t}{c}\right) + 61.4 \tag{12}$$

3. Simulation assumptions

In the analysis of sound pressure from wind turbine blade, generalized blade element momentum (BEM) method was used to compute the relative velocity field along the blade span. The outputs of BEM solver are relative velocity on the blade section, angle of attack, normal and tangential force coefficients on every section of blade which can be used to compute rotor loading forces and moments. The outputs from BEM solver are coupled to BPM noise prediction module for which, sound pressure level computations are done at a given wind speed, blade pitch angle and rotational speed of the machine.

In the BEM approach the total length of blade is discretized into several aerofoils at least 20 segments. Aerofoil can be assumed as half-infinite flat plate with finite thickness and aspect ratio. The flow over flat plate was assumed to be 2D incompressible and quasi uniform along the blade length which means that flow behavior does vary from one span station to another along the blade span. The overall shape of blade is approximated using selected aerofoils, viz. NACA 0012, NACA 6320 and

NACA 63215 while the turbulent boundary layer properties on suction and pressure side of aerofoils is computed from XFOIL module. The boundary layer data for aerofoil serve as input to the noise prediction module.

In the prediction of sound pressure levels, each blade segment is treated as a point source in near field and rotating blade as line source. In the far field sound prediction however, the rotor of turbine acts as point source when operating in a wind farm. Sound pressure level is thus calculated by logarithmic addition of individual sources relative to observer position. For the present simulation work, the receiver height was fixed at 2 m above the ground level and the source height was fixed at 80 m. The distance of the receiver location was set at 110 m, which is approximately the total turbine height ($HH + D/2$). This is in accordance with IEC 61400–11 regulations for measurements of acoustic emissions from wind turbines. HH is the hub height of turbine, and D is the rotor diameter in m.

A downwind scenario is considered as the worst case since sound waves bend in downward direction with respect to free stream wind and this results in amplification. Therefore, downwind receiver location is considered. The boundary conditions for the blade are Reynolds number, the angle of attack along the blade span. It is implemented to verify that blade element momentum (BEM) computed values do not exceed predefined threshold values as given in [6]. The blade pitch angle is set to 3.5° for sound pressure calculations and rotation speed for machine as 17 RPM.

3.1 Geometric model of turbine

For the assessment of trailing edge bluntness noise from horizontal axis wind turbine rotors, a geometric model for the blade has been developed using NuMAD software [18]. The software allows user to input the aerofoil data at every span wise location of the blade. **Table 1** shows the turbine design parameters along with orientation of rotor into wind.

Figure 2 shows the isometric (3D) model of the 38 m blade for the 2 MW wind turbine used for analyzing the trailing edge bluntness noise. Towards the inboard region, the airfoils have high thickness to chord ratio with at least 18% t/c as well as high camber. In the present study NACA 6320 airfoil data have been used with a

Parameter	Value
Cone angle	0°
Tilt angle	3°
Hub height	80 m
Blade Radius	38 m
Rotor speed	17 RPM
Max twist	13°
Max chord	3.22 m
Orientation	Upwind
No of blades	3
Rated power	2 MW

Table 1.
Turbine parameters for 2 MW machine.

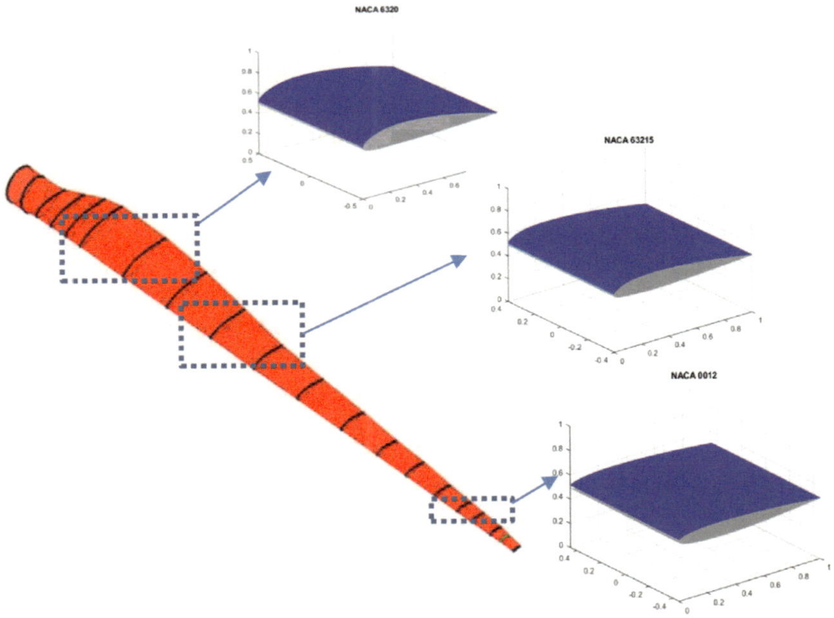

Figure 2.
Isometric (3D) model of wind turbine developed using NuMAD software showing the airfoil sections used near the blade root, mid-span, and tip of the blade [18].

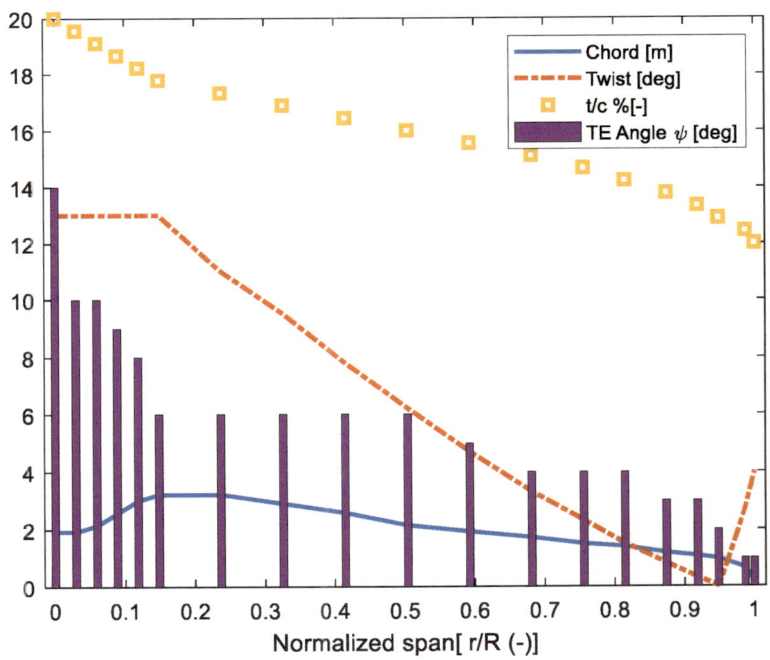

Figure 3.
Geometric properties shown along the normalized blade span for a wind turbine blade having a length of 38 m.

trailing edge slope angle of 14°. In the mid span region, the airfoils have moderate thickness to chord ratio. The geometric properties of the blade are depicted in **Figure 3**. It is evident that chord length and twist remain constant for root section which connects the blade to the rotor hub.

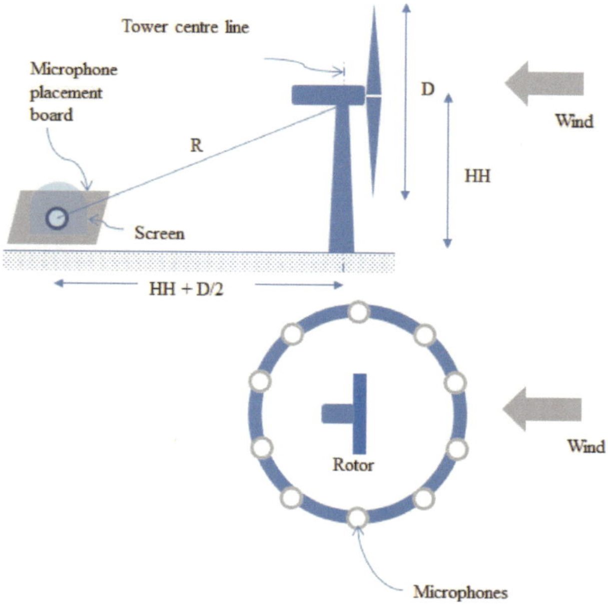

Figure 4.
Illustration of microphone position surrounding the source located in Centre as well as the microphone measurement distances and position according to IEC 61400–11 standards with respect to source.

3.2 IEC 61400-11 standard for measurement of sound level

The relative position of the receiver with respect to aerofoil coordinate system is shown in **Figure 4**. For a wind turbine blade, in addition to the turbulent boundary layer, wind speed and free stream mach number responsible for trailing edge vortex shedding noise, the sound levels also depend on blades pitch angle operation. Particularly for moderate pitch angles and at low or positive angle of attack, the boundary layer on the pressure side of aerofoil at leading edge shows laminar flow structure; however, the boundary layer on suction side remains mostly in turbulent state near the trailing edge. Further, it is important to note that such a type of noise mechanism is dominant in mid span region of blade where trailing edge thicknesses are high for which maximum Strouhal number is found to be 0.15. Below this value, the vortex shed from the trailing edge surface does not contribute significantly to the noise levels [8, 10].

4. Results and discussion

In this section we present results for the turbulent boundary layer vortex shedding noise from a 2 MW horizontal axis wind turbine blade using original BPM model predictions, modified by [9] and compare them with numerical computations implemented using the correction function for airfoil thickness. **Figure 5** illustrates the noise prediction from wind turbine blades using first order empirical methods which are based on the turbine geometric and operating parameters viz. rotor diameter, nominal power rating of machine and blade tip speed. Although such methods can predict sound power levels, they do not take account of the physical phenomenon responsible for the noise radiation from rotating blades at broadband frequencies and hence not reliable. Also, it can be said noise models proposed by [21–23] are simple algebraic functions that depend on nominal power

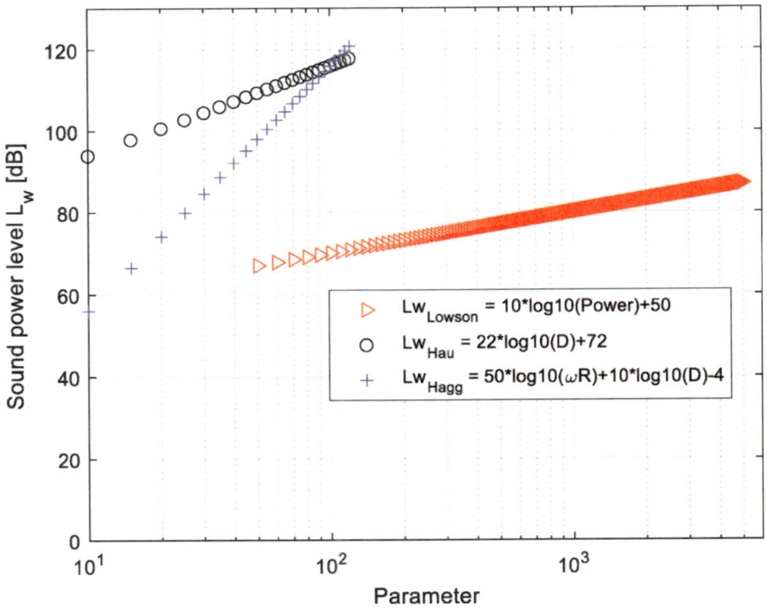

Figure 5.
Illustration of sound power level based on empirical relations proposed by [19–21].

Wind Turbine Model	dB(A)	Power (kW)	Hub Height (m)(m)	Rotor Diameter (m)
AN Bonus 600 kW/41	101.6	600	50	41
DeWind 41	99.6	500	40	41
DeWind 46	97.9	600	40	46
Enercon E-41	99	500	50	41
NORDTANK 500/41	103.2	500	50	41
SEEWIND 52–750-65	99	750	55	52
VESTAS V 66/1.65 MW	103	1650	60	66
Windtechnik-Nord 200/26	101	200	40	26

Table 2.
Sound power level, L_{wA} for utility scale commercial wind turbine models.

rating of the machine, rotor diameter and blade tip speed only. Sound power predictions from [22, 23] agree well for rotor diameters that range between 10 m and 100 m and thought to be less conservative compared to actual or measured data. Similarly, Lowson's empirical equation make use of only nominal power rating of machine, which implies that sound power level varies with size of machine. Hagg (1992) also developed a slightly more sophisticated model which can predict sound pressure level based on the axial thrust force coefficient, rotor swept area and the number of blades in machine along with empirical constants given in [21]. However, the model does not predict sound power levels for broadband frequency range of noise spectra. Some advanced noise prediction simulation software's developed by Siemens XNoise, NREL's NAFNoise are useful tools which can predict the noise levels for utility scale wind turbines. **Table 2** shows the measured sound power level (PWL) for some of commercial wind turbine models taken from SoundPLAN software.

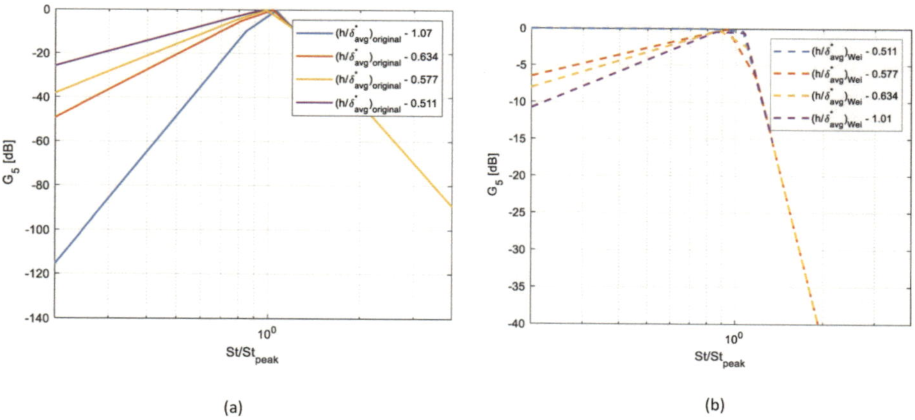

Figure 6.
Shape function, G_5, computed for different trailing edge bluntness thickness, h/δ^*_{avg} using (a) original BPM [6] (b) modified BPM by [9].

Figure 6(a) depicts the results for shape function, G_5 obtained from original BPM model. As the average boundary layer displacement thickness is reduced, the frequency of vortex shedding increased despite a change in the angle of attack and flow Mach number, *M* along the blade span. This difference can be attributed to solid angle inclusion in the original BPM model which considered trailing edge sloping angle, ψ as essential condition to vortex shedding phenomenon in addition to the trailing edge height, flow Mach number and Reynolds number.

From **Figure 6(b)** the shape function G_5 modified by [9] has been computed for trailing edge height to average boundary layer displacement thickness ratios, h/δ^*_{avg} between 0.51 and 1.01. The function showed a linear change in amplitude, dB for all Strouhal numbers of ratios between 0 and 1. As the peak Strouhal number, St_{peak} is increased, one can notice that tonal peak for trailing edge bluntness was found to be increasing. This effect was also observed with numerical CAA results obtained by [9] in their study for NACA 0012 and NACA 63–418 airfoil which have ~3% camber and maximum thickness of 18%. It is important to note that CAA computations such as large eddy simulation (LES) can predict the acoustic radiation from airfoils by solving for the largest scales of turbulent flows and approximating the small scale motions. In contrast to the semi-empirical BPM model, the sound pressure level near the surface can be computed by solving the 2D-Navier–Stokes (N-S) equations that are coupled to advanced turbulence models and high accuracy computational grid schemes suitable for acoustic pressure computations [14]. Similarly, the A-weighted 1/3rd octave band tonal noise spectra has been computed at wind speed of 6 m/s, 14 RPM having a blade pitch of 3.5°.

Figure 7(a)–(d) demonstrates the contour plot of peak Strouhal number, plotted along the blade span for various blade azimuth angles in rotor plane and for different trailing edge thicknesses computed at wind speed of 8 m/s [19]. The maximum values can be observed between 0.1 r/R and 0.75 r/R along the blade span where the thickness to chord ratio is high when the blade azimuth angle is at 300°. With increasing trailing edge thicknesses, the peak Strouhal number kept increasing from 0.13 to 0.2. This also signifies shape function, G_5 have high tonal peaks demonstrating influence of trailing edge vortex shedding from blade caused due to change in the trailing edge thicknesses.

On the other hand, the original BPM model showed a strong tonal peak effect in noise spectrum at 12% r/R where the thickness to chord ratio is found increasing. **Figure 8** shows the computed values for overall A-weighted 1/3rd octave band sound power level for 2 MW turbine, turbulent boundary layer trailing edge noise

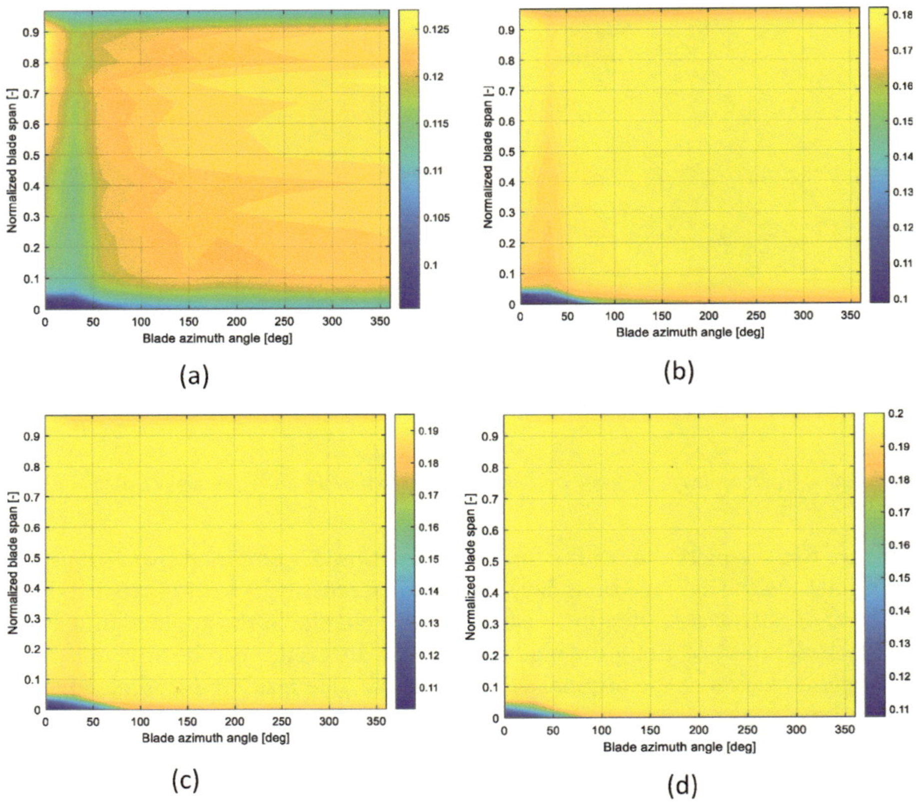

Figure 7.
Peak Strouhal number, St_{peak}''', along normalized blade span and blade azimuth angles at U = 8 m/s for TE thicknesses (a) 0.1% chord (b) 0.5% chord (c) 1% chord (d) 1.5% chord.

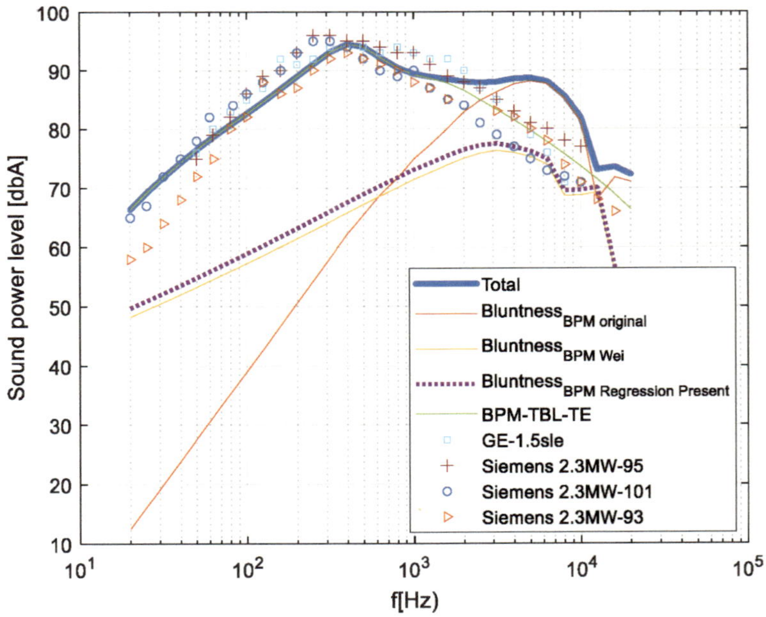

Figure 8.
Comparison of trailing edge bluntness noise using present approach to those predicted by BPM original, BPM modified [9]) and its validation with experimental data from Siemens SWT 2.3 MW and GE 1.5sle turbines at wind speed of 8 m/s for trailing edge thickness of 0.1% chord.

(TBL-TE) as well as trailing edge bluntness noise using original BPM model. All the computations were done in MATLAB 2020b software, for a wind speed of 8 m/s at blade pitch angle of 3.5° and trailing edge thickness taken ~0.1% chord length. Further, modified BPM model by [9] for trailing edge bluntness noise has also been computed to compare the actual results with present method. The present method focused on the regression approach for thickness correction along the blade span. It can be noted that present results produced similar trailing edge noise characteristics except for the noise peak change found at 10 kHz in the noise spectra. Also, one can observe that current approach for thickness correction leads to better agreement of the trailing edge bluntness peak with experiment data obtained from GE 1.5sle rather than Siemens 2.3 MW-101, Siemens 2.3 MW-95 and Siemens 2.3 MW-93 turbines. On the contrary, the trailing edge bluntness peak from original BPM model showed a broad hump which do not agree well with experiment validation data for turbines. For frequencies below 1 kHz, the turbulent boundary layer trailing edge noise dominates with a peak value of 96dBA. The trailing edge bluntness noise tonal peak computed from the original BPM model was found to be ~89 dBA. The peak trailing edge bluntness noise level for modified BPM by [9] was found to be 78 dBA near 8 kHz which agreed well with experiment data. The present computations for modified BPM showed an increase of 2 dBA for frequency range of 20 Hz and 6 kHz, but reached almost same values for frequencies, f > 6 kHz.

In this section we present results for the turbulent boundary layer vortex shedding noise from a 2 MW horizontal axis wind turbine blade using original BPM model predictions and compare them with OSPL (overall sound power level) experiment data obtained for GE 1.5sle, Siemens SWT 2.3 MW machines. All the computations were done in MATLAB 2020b software, for a wind speed of 7 m/s and 10 m/s at blade pitch angle of 3.5° and trailing edge thickness taken ~0.1% chord length. **Figure 9** shows the Strouhal number, St''' computed in terms of displacement thickness, δ^*, for wind speeds of 7 m/s and 10 m/s respectively. The maximum value for St''' was found to be 2.2 and 4.16 for wind speeds of 7 m/s and 10 m/s at frequency f ~ 10 kHz, where the turbulent boundary layer trailing edge bluntness noise produces peak tonal amplitude.

Figure 10(a) and **(b)** shows the trailing edge bluntness peak from original BPM model as a broad band hump that agrees well within 5% of experiment validation data for GE 1.5sle, Siemens 2.3 MW turbines for wind speed of 7 m/s and 10 m/s at

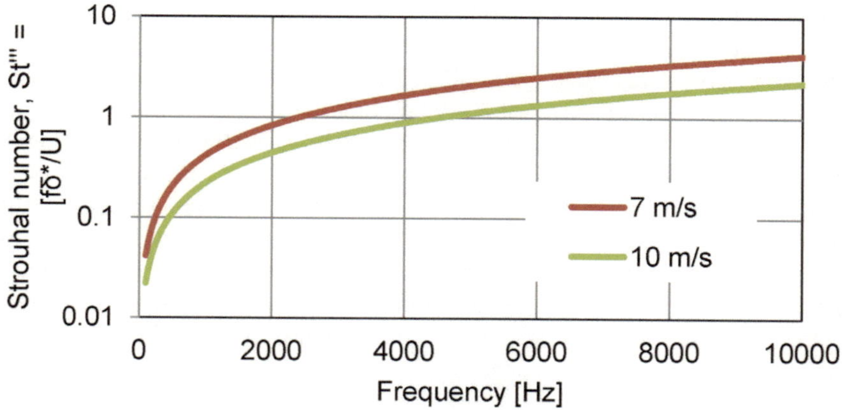

Figure 9.
Illustration of Strouhal number, St''' as function of displacement thickness, δ^*, at wind speeds of 7 m/s and 10 m/s.

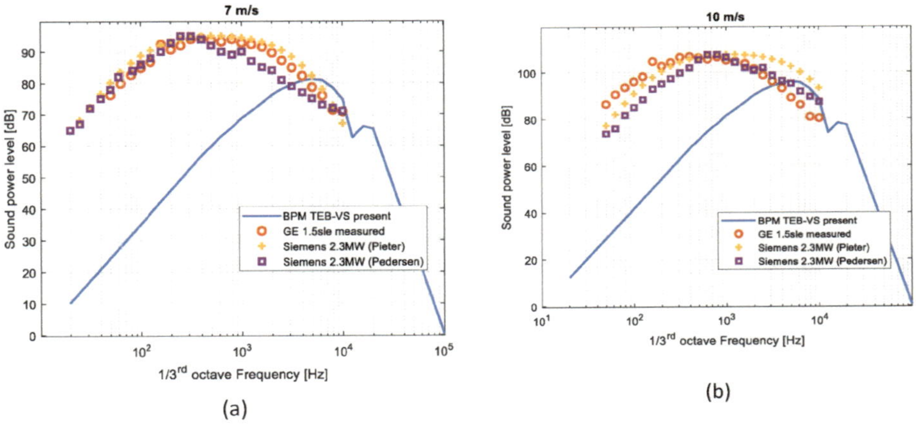

Figure 10.
Validation of the computed BPM-turbulent boundary layer vortex shedding noise (TEB-VS), for a blade length of 38 m, 2 MW turbine having trailing edge thickness of 0.1%c with OSPL measured data of GE-1.5sle, Siemens 2.3 MW (Pieter), Siemens 2.3 MW (Pedersen) of blade length 47 m at two different wind speeds of (a) 7 m/s (b) 10 m/s.

trailing edge thickness of 0.1% local blade chord length. For frequencies below 1 kHz, the turbulent boundary layer trailing edge noise dominates with a peak value of 96 dB that is obtained using measured data of experiment turbines. The tonal peak of trailing edge bluntness noise computed from the original BPM model was found to be ~82 dB for wind speed of 7 m/s and 96 dB for wind speed of 10 m/s.

From **Figure 11(a)** and **(b)** one can notice the computed turbulent boundary layer trailing edge bluntness noise level using BPM model shows peaks that shift closer to frequencies, f ~ 5 kHz and reach an amplitude values of 97 dB and 115 dB respectively. It must be noted that when the trailing edge thickness or heights are increased to 0.5% of local blade chord length, a difference of 15 dB was found for wind speed of 7 m/s while a difference of 10 dB was obtained for wind speed of 10 m/s. Further, from **Figure 12** it is evident that the difference in the sound power levels between 7 m/s and 10 m/s continued to increase by a maximum value of 15 dB for frequencies, f < 200 Hz when the trailing edge thicknesses are 0.1% and 0.5% of local blade chord length respectively. However, for frequencies, f > 200 Hz a noise reduction of 17 dB was observed when the trailing edge thickness was 0.5% local chord length.

Figure 13 shows the measured and computed sound power level, L_{wA} for wind speeds between 4 m/s and 10 m/s. The experiment data for Vestas V82 and GE 1.5sle turbines have source heights of 80 m and blade lengths of 40 m which are nearly same as present investigated 2 MW turbine. This data is obtained for one of Vestas V82 and GE 1.5sle turbines from Jericho Rise operating wind farm located in US state of New York [20]. The results demonstrate that for wind speeds lesser than 7 m/s both experiment noise data for Vestas 82 and GE 1.5sle agree closely with each other within 1%. However, from 7 m/s to 10 m/s the sound power level remained constant which implies that there is no influence of wind speeds on sound levels which contradicts the BPM model predictions as the model is strongly dependent on Mach number. This suggests that turbines are deliberately controlled above certain wind speeds in order to regulate power. Further, the model simulated values for the present case of 2 MW turbine also agree closely with experiment data of both turbines with a peak difference of 5dBA at wind speed of 6 m/s. This shows that model can predic the sound levels accurately and reliably be used for the noise assessment of wind turbines.

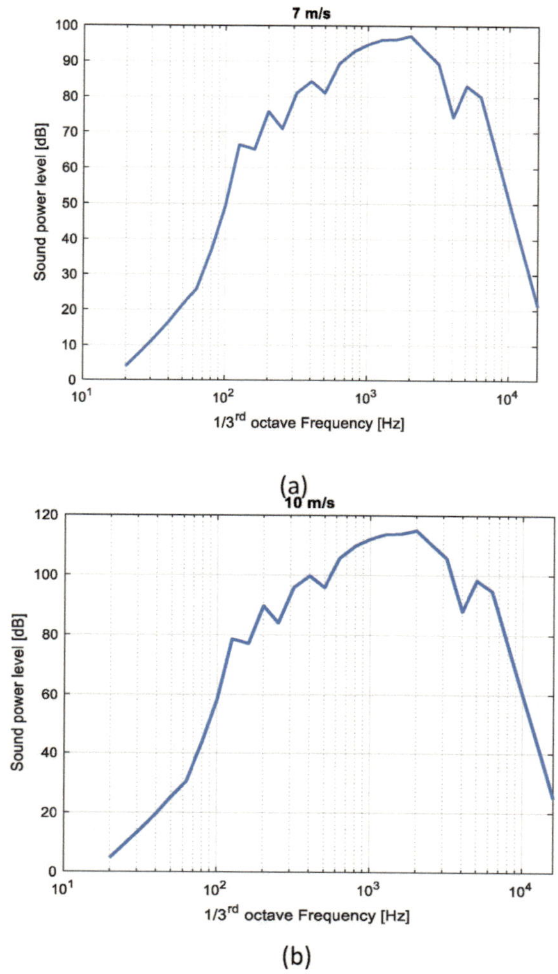

Figure 11.
Computed turbulent boundary layer vortex shedding noise (TEB-VS) for a blade length of 38 m, 2 MW turbine using trailing edge thickness of 0.5%c at two different wind speeds (a) 7 m/s (b) 10 m/s.

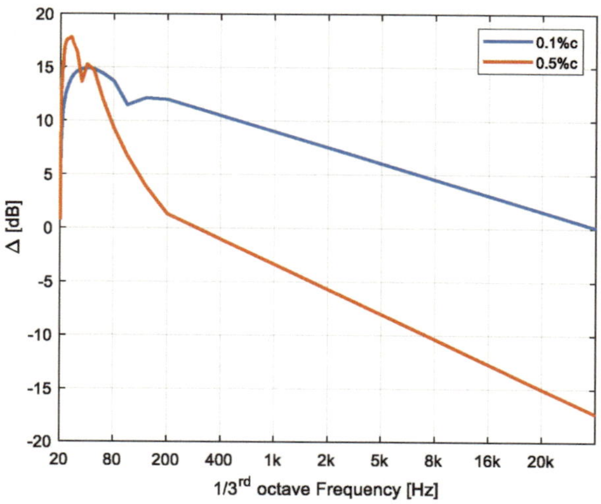

Figure 12.
Computed difference, Δ dB, of the turbulent boundary layer vortex shedding noise level (TEB-VS) between wind speeds 7 m/s and 10 m/s, at trailing edge thicknesses of 0.1%c and 0.5%c.

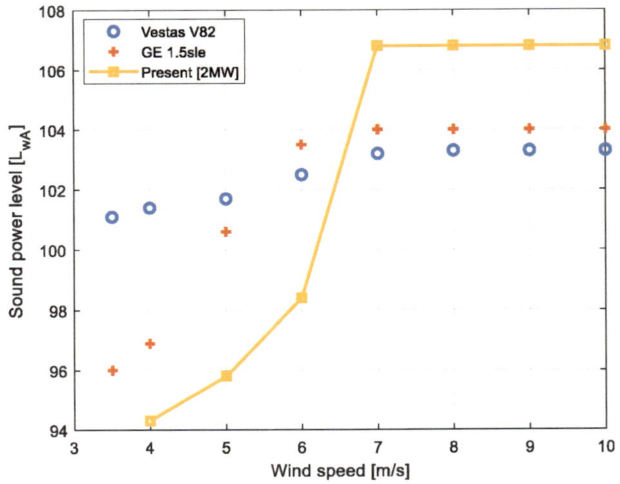

Figure 13.
Sound power levels computed for various wind speeds, for present 2 MW turbine, 38 m blade length compared to experiment data for Vestas V82, GE 1.5sle at same hub heights.

5. Conclusion

A computational analysis of trailing edge bluntness vortex shedding noise for 2 MW horizontal axis turbine was performed for trailing edge thicknesses of 0.1% and 0.5% local chord using original BPM model. The original BPM results for trailing edge bluntness noise showed that for a trailing edge thickness of 0.1% and 0.5% chord length the effect on sound power level was found to be ~83 dB, 92 dB and 95 dB and 115 dB at wind speeds of 7 m/s and 10 m/s respectively. At 10 kHz region turbulent boundary layer vortex shedding noise masks all other self-noise mechanisms. Finally, the existing overall sound power level (OSPL) experimental data showed very good agreement with simulated outputs for the trailing edge bluntness noise at wind speeds 7 m/s 8 m/s and 10 m/s respectively. The original BPM results for trailing edge bluntness noise showed that for a trailing edge thickness of 0.1% chord the effect on 1/3rd octave overall A weighted sound level was found as a peak hump with an amplitude of ~90 dBA near 10 kHz region and masks all other self-noise mechanisms. The modified BPM results for trailing edge bluntness noise also showed a sharp peak instead of a hump but the amplitude reduced by ~15 dBA at 8 kHz in noise spectra. The new thickness correction function predicted the peak amplitude of trailing edge bluntness more accurately compared to original and modified BPM.

Conflict of interest

Authors declare no conflict of interest for the present work.

Nomenclature

M	Mach number
h/δ^*	Trailing edge bluntness ratio
δ^*	Boundary layer displacement thickness, mm
L	Length of span segment, m

r_e	Distance between the source and receiver position, m
G_4, G_5	Shape functions
θ, ϕ	Directivity angles between the source and receiver line
SPL	Sound power level, sound pressure level, dB
D_h	High frequency directivity
Ψ	Trailing edge angle, degree
K_o	Empirical constant
r	Local radius, m
R	Blade radius, m
t/c	thickness to chord ratio
L_{wA}	A -weighted sound power level
St'''	Strouhal number
HH	Hub height, m
D	Rotor diameter, m
St_{peak}'''	Peak Strouhal number
δ^*_{avg}	Average boundary layer displacement thickness, mm
M_c	Critical Mach number
f	Frequency, Hz
U	Free stream wind velocity, wind speed, m/s
c	Chord length

Author details

Satya Prasad Maddula[1], Vasishta Bhargava Nukala[2]*,
Swamy Naidu Neigapula Venkata[3], Chinmaya Prasad Padhy[4] and Rahul Samala[5]

1 Department of Aerospace Engineering, GITAM (Deemed to be University), Hyderabad, India

2 Department of Mechanical Engineering, Sreyas Institute of Engineering and Technology, Hyderabad, India

3 Department of Mechanical Engineering, NIT Raipur, Raipur, India

4 Department of Mechanical Engineering, GITAM (Deemed to be University), Hyderabad, India

5 Department of Applied Mechanics, Indian Institute of Technology, Chennai, India

*Address all correspondence to: vasishtab@gmail.com

References

[1] Bastasch M, van Dam J, Sondergaard B, Rogers A. Wind Turbine Noise – An Overview. Journal of Canadian Acoustical Association. 2006;**34**(2):7-16

[2] Geyer T, Sarradj E, Fritzsche C. Porous Aerofoils: Noise Reduction and Boundary Layer Effects. International Journal of Aeroacoustics. 2010;**9**(6): 787-820. DOI: 10.1260/1475-472X.9.6.787

[3] Tang H, Lei Y, Fu Y. Noise reduction mechanisms of an airfoil with trailing edge serrations at low Mach number. *MDPI: Applied science*; 2019 https://doi.org/10.3390/app9183784

[4] Kingan, K.M.*Aero-Acoustic Noise Produced by an Aerofoil* [Doctoral thesis]. Christchurch: University of Canterbury, New Zealand, 2005, p. 448. Available from: https://ir.canterbury.ac.nz/handle/10092/6596

[5] Brooks. T.F and Hodgson. T.H, Trailing Edge Noise Prediction from Measured Surface Pressures. *Journal of Sound and Vibration,* vol 78, p 69-117. 1981, DOI: https://doi.org/10.1016/S0022-460X(81)80158-7

[6] Brooks, T.F., Pope, D.S. and Marcolini, M.A, *Airfoil Self Noise and Prediction.* NASA reference publication 1218, 1989, Available from: https://ntrs.nasa.gov/archive/nasa/casi.ntrs.nasa.gov/19890016302.pdf

[7] Zhu WJ. *Modelling of Noise from Wind Turbines* [Master thesis]. Lyngby: Department of Wind Energy, Technical University of Denmark; 2004

[8] Moriarty, P. and Migliore, P.(2003) *Semi Empirical Aero-Acoustic Noise Prediction Code for Wind Turbines* [Technical report], 2003, 39 p. Available from:http://citeseerx.ist.psu.edu/viewdoc/download?doi=10.1.1.197.1153&rep=rep1&type=pdf

[9] Wei, Z. J; Shen, W.Z; Sorensen, J.N; Leloudas, G (2016) Improvement of airfoil trailing edge bluntness noise model, *Advances in mechanical engineering* , 2016, vol 8, p 1-12, DOI: 10.1177/1687814016629343

[10] Grosveld FW. Prediction of Broadband Noise from Horizontal Axis Wind Turbines. Journal of Propulsion and Power. 1985;**1**(4):292-299 ISSN 0748-4658

[11] Howe. M. S *Theory of Vortex Sound* Cambridge University Press. 2003, DOI: https://doi.org/10.1017/CBO9780511755491

[12] Doolan C, Moreau DJ, Arcondoulis E, Albarracin C. Trailing Edge Noise Production. Prediction and Control. *New Zealand Acoustics*. 2012; **25**(3):22-29

[13] Blandeau, V.P and Joseph, P.F, Validity of Amiet's Model for Propeller Trailing Edge Noise. AIAA Journal, 2011, vol. 49, no 5, p. 1057-1066. DOI 10.2514/1.J050765.

[14] Dijkstra, P, *Rotor Noise and Aero-Acoustic Optimization of Wind Turbine Aerofoils* [Master Thesis]. Delft: Delft University of Technology. 131 p. 2015.

[15] Leloudas, G, *Optimization of Wind Turbines with Respect to Noise* [Master Thesis]. Lyngby: Technical University of Denmark, p. 66, 2006.

[16] Blake, W.K, *Aero-Hydro-Acoustics for Ships Volume II*, [Technical Report] 1984. Available from: https://apps.dtic.mil/dtic/tr/fulltext/u2/a150672.pdf

[17] Moreau, D.J, Brooks, L. A. and Doolan, C, Flat Plate Self-Noise Reduction at Low to Moderate Reynolds Number with Trailing Edge Serrations. In *Proceedings of Acoustics.* Gold Coast: Australia. 2011, Available from: https://

pdfs.semanticscholar.org/74fa/0f9f258c0d5c5a1f846abb244c6d7505b213.pdf

[18] NuMAD software, Version: NuMADexe_130403_PCWIN64. Available from: https://energy.sandia.gov/energy/renewable-energy/wind-power/rotor-innovation/numerical-manufacturing-and-design-tool-numad.

[19] Bhargava V, Maddula SP, Samala R. Prediction of vortex induced aerodynamic noise from wind turbine blades. Advances in Military Technology. 2019;**14**(p):245-261. DOI: 10.3849/aimt.01295

[20] Appendix 1, Environmental sound survey and noise impact assessment, *Draft Environmental Impact Assessment, Jericho Rise Wind Farm LLC. TetraTech EC Inc, Boston, USA, 2008.*

[21] Lowson, M. V., Assessment and Prediction of Wind Turbine Noise, Flow Solutions Report 92/19, ETSU W/13/00284/REP, pp. 1-59, 1992.

[22] Hau E, Langenbrinck J. Palz, W. Springer-Verlag, Berlin: WEGA Large wind turbines; 1993. pp. 1-143

[23] Hagg, F., van der Borg., N. J. C.M Bruggeman, J. C.; et al. Definite Aero-Geluidonderzoek Twin," Stork Product Engineering B.V., SPE 92- 025. 1992.